2012

Next Generation
Experiments to Measure
the Neutron Lifetime
Workshop

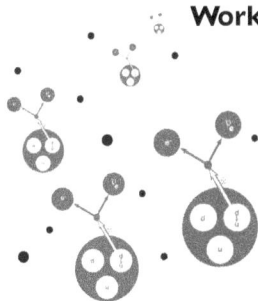

2012

Next Generation Experiments to Measure the Neutron Lifetime Workshop

Santa Fe, New Mexico
9 – 10 November 2012

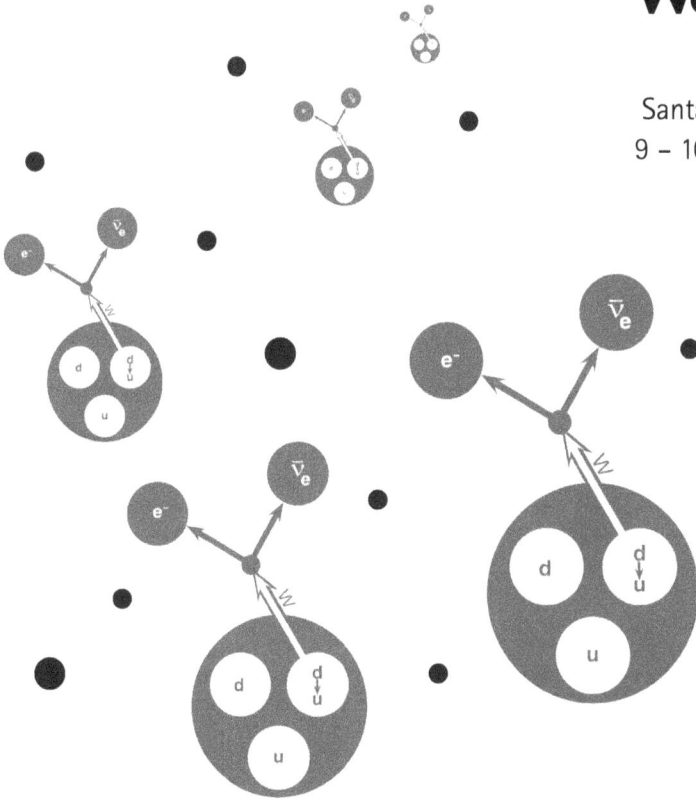

Editor

Susan J. Seestrom
Los Alamos National Laboratory, USA

World Scientific

NEW JERSEY · LONDON · SINGAPORE · BEIJING · SHANGHAI · HONG KONG · TAIPEI · CHENNAI

Published by

World Scientific Publishing Co. Pte. Ltd.

5 Toh Tuck Link, Singapore 596224

USA office: 27 Warren Street, Suite 401-402, Hackensack, NJ 07601

UK office: 57 Shelton Street, Covent Garden, London WC2H 9HE

British Library Cataloguing-in-Publication Data
A catalogue record for this book is available from the British Library.

NEXT GENERATION EXPERIMENTS TO MEASURE THE NEUTRON LIFETIME

ISBN 978-981-4571-66-1

In-house Editor: Rhaimie Wahap

Preface

This workshop was motivated by a desire to bring together the vibrant and growing community that is using low energy neutrons to study neutron decay. The present confusing state of knowledge of the neutron lifetime calls out to be addressed by new experiments, and recent technical developments provide great opportunities for improved results. Most of the groups working in this area world-wide were represented.

The workshop ended prematurely because of the tragic death of Stuart Freedman while he was in Santa Fe to serve on the Expert Panel providing the summary of this workshop. Stuart has been an incomparable figure in nuclear physics in a wide range of areas — always bringing his sharp intellect and sense of humor to the benefit of our field. Stuart served as a colleague, mentor, and friend to many of us working on fundamental symmetries and he will be deeply missed.

We dedicate this volume to Stuart Freedman.

Susan J. Seestrom
Editor
(for the organizing committee of J. David Bowman, Paul Huffman, Akira Konaka, Chen-Yu Liu, Pieter Mumm, Alexander Saunders, and Susan J. Seestrom)

Contents

Neutron Lifetime Theory

W. J. MARCIANO

Brookhaven National Laboratory
P.O. Box 5000
Upton, NY 11973-5000, USA
marciano@bnl.gov

Standard Model electroweak radiative corrections to the neutron lifetime, τ_n, are reviewed. A resulting "Master Relation" connecting τ_n, g_A and V_{ud} is described. The utility of measuring τ_n and g_A to $\pm 0.01\%$ as a means of definitively testing CKM unitarity is discussed.

Keywords: Neutron lifetime

The neutron lifetime, τ_n, is of fundamental importance in low energy nuclear physics.[1–3] Its value directly impacts primordial nucleosynthesis constraints on the "effective" number of neutrino species. In addition, τ_n, used together with the muon lifetime and CKM matrix element, V_{ud}, determines $g_A \equiv G_A/G_V$ at $q^2 \simeq 0$ with very high precision.[4] That non-perturbative QCD parameter has changed dramatically over the years, increasing[1] from $g_A \simeq 1.18 \rightarrow 1.13 \rightarrow 1.25 \rightarrow 1.27$ over time. Its value also impacts primordial nucleosyntehesis, as well as: solar and reactor neutrino flux calculations, the Goldberger-Treiman relation, muon capture,[5] the Bjorken Sum Rule etc., all topics of current interest. In this talk, however, I will not discuss those applications. Instead, I focus on the utility of τ_n and g_A, measured to $\pm 0.01\%$ as a means of determining the CKM matrix element, V_{ud}, with high precision, $\simeq \pm 0.02\%$, independent of nuclear uncertainties. $|V_{ud}|$, used in conjunction with $|V_{us}|$ and $|V_{ub}|$ allows for a definitive test of CKM unitarity via the relationship[6]

$$|V_{ud}|^2 + |V_{us}|^2 + |V_{ub}|^2 = 1 \qquad (1)$$

An "apparent" experimental deviation from 1 would be indicative of "new physics".

The current best value of V_{ud}, obtained from an average of super-allowed $(0^+ \rightarrow 0^+)$ nuclear beta decays over a broad range of nuclei[7]

$$|V_{ud}| = 0.97425(11)_{\text{nuclear}}(19)_{\text{RC}} \qquad (2)$$

used together with $|V_{us}| = 0.2253(9)$ obtained from K decays[8] and the rather negligible $|V_{ub}|^2 = 1.5 \times 10^{-5}$ leads to

$$|V_{ud}|^2 + |V_{us}|^2 + |V_{ub}|^2 = 0.9999(4)_{V_{ud}}(4)_{V_{us}} \qquad (3)$$

in excellent agreement with unitarity and showing no sign of "new physics". Eq. (3) has been used to constrain:[9] heavy quark or lepton mixing, supersymmetry, charged Higgs scalars, excited gauge bosons etc. However, despite the small nuclear error in eq. (2) and continuing efforts to validate it, there remain questions regarding the magnitude of nuclear Coulombic isospin breaking corrections.[7] Qualitative estimates[10–13] suggest the possibility of a smaller $|V_{ud}|$. For that reason, it would be useful to have a competitive determination of $|V_{ud}|$ via neutron beta decay which is independent of nuclear physics uncertainties.

To confront $|V_{ud}|$ with $\pm 0.02\%$ (or better) precision in neutron decay, requires careful consideration of electroweak radiative corrections to τ_n and g_A. The standard approach, pioneered by Sirlin[14] is to compare the neutron lifetime, τ_n, with the precisely measured muon lifetime[15]

$$\tau_{\mu^+} = 2.1969803(22) \times 10^{-6} \text{ sec} \qquad (4)$$

where the full 1 & 2 loop radiative corrections are known. With their inclusion, the muon lifetime then leads to a very accurate Fermi constant determination[15]

$$G_\mu = 1.1663787(6) \times 10^{-5} \text{ GeV}^{-2} \qquad (5)$$

which is used to normalize all other weak interaction amplitudes. In the case of the neutron decay rate (inverse lifetime), one finds[4]

$$\tau_n^{-1} = f G_\mu^2 |V_{ud}|^2 m_e^5 (1 + 3g_A^2)(1 + RC)/2\pi^3 \qquad (6)$$

where

$$f = 1.6887 \qquad (7)$$

is a phase-space factor that includes:[16] Fermi function effects, proton recoil, nucleon size corrections etc. RC in that expression represents the Standard Model (SM) electroweak radiative corrections to τ_n not already absorbed into G_μ. The quantity $g_A \equiv G_A/G_V$ is essentially defined by eq. (6), but determined independently using neutron decay final state asymmetries.[1]

The RC in eq. (6) are finite and calculable. At the 1 loop level[14]

$$RC = \frac{\alpha}{2\pi} \left[\langle g(E_{\max}) \rangle + 3\ell n(m_Z/m_p) + \ell n(m_Z/m_A) + 2C + A_{\text{QCD}} \right] \quad (8)$$

where $g(E)$ is the Universal Sirlin Function[17] which is averaged over the neutron decay spectrum to give

$$\frac{\alpha}{2\pi} \langle g(E_{\max}) \rangle = 0.015056. \quad (9)$$

The $\ell n m_Z$ terms are short-distance vector and axial-vector loop corrections from the γW box diagrams with $m_A \simeq 1.2$ GeV a long-short distance matching scale. The long distance axial $C \simeq 0.8 g_A(\mu_n + \mu_p) \simeq 0.9$ and perturbative QCD corrections to the short-distance axial-current contribution, $A_{\text{QCD}} \simeq -0.34$. Most of the RC uncertainty resides in the rough loop matching condition parametrized by m_A. Higher order loop corrections give an additional $+0.0013(1)$ contribution to eq. (8) such that in 2004, one found,[4] $RC \simeq 0.0390(8)$. Later,[18] smoother matching conditions were employed along with additional perturbative QCD corrections to improve RC to

$$1 + RC = 1.03886(39) \quad (10)$$

where the uncertainty was reduced by about $1/2$.

Employing eq. (10) in eq. (2) leads to the master neutron decay relation

$$|V_{ud}|^2 = \frac{4908.7(1.9) \text{ sec}}{\tau_n(1 + 3g_A^2)} \quad (11)$$

connecting $|V_{ud}|$, τ_n and g_A. Future more aggressive treatment of the error budget together with the use of dispersion relations could reduce the error in eq. (11) by about a factor of $2/3$; however, beyond that will be difficult.

Planned experimental programs will aim for $\pm 0.01\%$ determinations of τ_n and g_A, leading to $\Delta V_{ud}/V_{ud} \simeq \pm 0.0002$ with the primary uncertainty coming from the RC error in eq. (10). At that level of precision, neutron decay will be competitive with super-allowed nuclear beta decays for determining $|V_{ud}|$, but without the nuclear uncertainty. Will it confirm or challenge the unitarity constraint in eq. (3)?

Instead of using the "Master Relation" in eq. (11) to determine $|V_{ud}|$, one can employ $|V_{ud}| = 0.97425(11)_{\text{nucl.}}(19)_{RC}$ from superallowed nuclear decays (or similarly from $|V_{us}|$ + unitarity), in which case eq. (11) gives

$$\tau_n(1 + 3g_A^2) = 5171.6(1.2) \text{ sec} \qquad (12)$$

Consistency with the current PDG average, $\tau_n^{\text{ave}} = 880.1(1.1)$ sec then requires

$$g_A = 1.2749(10)_{\tau_n}(2)_{\text{nucl.}} \qquad (13)$$

So, one expects g_A^{exp} to be heading higher, toward 1.275 if the central τ_n^{ave} remains fixed and unitarity is respected. That expectation is consistent with the recent single best measurement from Perkeo II,[19] $g_A = 1.2755(13)$.

The next few years should be very exciting for neutron decay studies. We will watch τ_n^{exp} and g_A^{exp} advance forward towards $\pm 0.01\%$ accuracy. Where will they wind up? Will they satisfy the unitary relationship in eq. (12) or will they present us with an inconsistency suggesting "new physics"?

References

1. H. Abele, *Prog. Part. Phys.* **60**, 1 (2008).
2. D. Dubbers and Schmidt, M., *Rev. Mod. Phys.* **83**, 1111 (2011).
3. F. Wietfeldt and G. Greene, *Rev. Mod. Phys.* **83**, 1173 (2011).
4. A. Czarnecki, W. Marciano and A. Sirlin, A., *Phys. Rev.* **D70**, 093006 (2004).
5. A. Czarnecki, W. Marciano and A. Sirlin, *Phys. Rev. Lett.* **99**, 032003 (2007).
6. W. Marciano and A. Sirlin, *Phys. Rev. Lett.* **56**, 22 (1986).
7. J. Hardy and I. Towner, I., *Phys. Rev.* **C79**, 055502 (2009).
8. PDG Collaboration, J. Beringer, *et al.*, *Phys. Rev.* **D86**, 010001 (2012).
9. W. Marciano, *Journal of Phys.* **312**, 102002 (2011).
10. G. Miller and A. Schwenk, A., *Phys. Rev.* **C78**, 035501 (2008).
11. G. Miller and A. Schwenk, *Phys. Rev.* **C80**, 064319 (2009).
12. N. Auerbach, *Phys. Rev.* **C79**, 035502 (2009).
13. H. Liang, N. Van Giai, and J. Meng, *Phys. Rev.* **C79** 064316 (2009).
14. A. Sirlin, *Rev. Mod. Phys.* **50**, 573 (1978).
15. MuLan Collaboration, D. Webber, *et al.*, *Phys. Rev. Lett.* **106**, 041803 (2011).
16. D. Wilkinson, *Nucl. Phys.* **A377**, 474 (1982).
17. A. Sirlin, *Phys. Rev.* **164**, 1767 (1967).
18. W. Marciano and A. Sirlin, *Phys. Rev. Lett.* **96**, 032002 (2006).
19. S. Capitani, *et al.*, *Phys. Rev.* **D86**, 074502 (2012).

Neutron Lifetime Experiments Using the Beam Method: Past, Present, and Future

F. E. WIETFELDT

Department of Physics, Tulane University,
New Orleans, LA 70118, USA
few@tulane.edu

The oldest method for measuring the neutron lifetime combines the rate of neutron decay in a slow neutron beam, found by counting the final state charged particles, with absolute determinations of the neutron flux and the effective decay volume to obtain the specific activity of the beam and hence the radioactive decay constant. This is known as the "beam method." The most recent beam neutron lifetime experiment[1] achieved an experimental uncertainty of 3.4 s, dominated by systematic limitations of the neutron flux measurement. Recent ultracold neutron "bottle" experiments have obtained uncertainties at or just below the 1 s level. The effort to reduce the neutron lifetime uncertainty to 0.1 s will benefit from both approaches because, while they are both challenged by difficult systematic effects at that level, the nature of these effects is very different for the two methods. We present a brief review of the principles and history of the beam method and outline a path for future beam experiments to improve the neutron lifetime uncertainty to 0.1 s and below.

Keywords: Neutron decay, beta decay

1. Principles of the Beam Method

The basic features of a beam neutron lifetime measurement are shown in Figure 1. A collimated slow (thermal or cold) neutron beam, produced by a research reactor or spallation neutron source and delivered by a critically reflecting neutron guide, is passed through a well known decay volume. The typical neutron beam velocity is about 1000 m/s and the decay region is of order 1 m in length, so with a lifetime of 880 s the probability for an individual neutron to decay in the region is only $\sim 10^{-6}$. Nevertheless at a modern high flux source a neutron decay rate of over 100 s^{-1} can be achieved. The decay region is viewed by one or more charged particle detectors that count the final state electrons or protons (or both) from neutron beta decay. The earliest beam lifetime experiments detected both in coincidence in order to reduce backgrounds. In more recent experiments only the protons have been counted to more precisely fix the effective decay volume and counting efficiency. Finally, the neutron beam passes through a thin absorbing foil that continuously monitors the neutron flux.

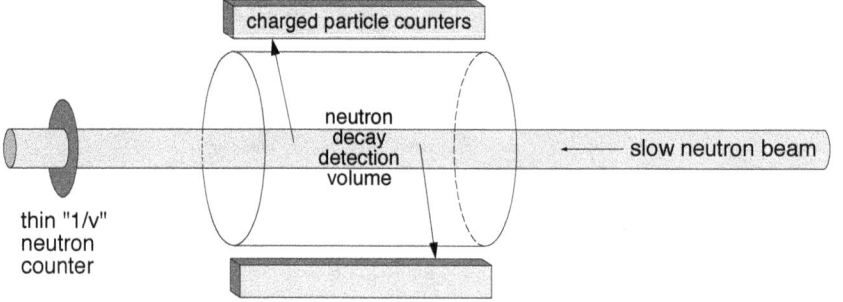

Fig. 1.   The basic layout of a beam neutron lifetime experiment.

The neutron decay rate Γ in the decay region is given by the differential equation that governs radioactive decay:

$$\Gamma = -\frac{dN}{dt} = \frac{N}{\tau_n} \tag{1}$$

where τ_n is the neutron lifetime and

$$N = \rho_n V_{\text{det}} = \left(\frac{\phi_n}{v_n}\right) A_{\text{beam}} L_{\text{det}} \tag{2}$$

is the total number of neutrons in the decay region. Here ρ_n is the average neutron density which can be written as the average neutron flux ϕ_n divided by the neutron velocity v_n, and V_{det} is the detection volume. If we assume the latter is large enough to encompass the full cross section of the beam A_{beam} then we replace V_{det} with the product A_{beam} times the length of the decay region L_{det}. Combining Eqs. 1 and 2 we have an expression for the neutron lifetime appropriate for a monochromatic (single velocity v_n) neutron beam:

$$\tau_n = \frac{A_{\text{beam}} L_{\text{det}}}{\Gamma} \left(\frac{\phi_n}{v_n}\right). \tag{3}$$

In practice a "white" neutron beam with a broad velocity spectrum is used so Eq. 3 requires an integral of the spectral flux $\phi_n(v)$:

$$\tau_n = \frac{A_{\text{beam}} L_{\text{det}}}{\Gamma} \int \frac{\phi_n(v)}{v} dv. \qquad (4)$$

The neutron decay detector counts the decay products (*e.g.* protons) with efficiency ε_p so the observed neutron decay rate R_p is

$$R_p = \varepsilon_p \Gamma \qquad (5)$$

The thin foil flux monitor is made of a neutron absorber such as ^{6}Li or ^{10}B whose absorption cross section obeys the "1/v" law to sufficient precision (to within 10^{-5}):

$$\sigma_{\text{abs}} = \sigma_{\text{th}} \frac{v_{\text{th}}}{v} \qquad (6)$$

where v_{th} is the reference thermal neutron velocity of 2200 m/s. The thin foil is viewed by a set of particle detectors that count the reaction products from neutrons absorbed in the foil. The thermal flux monitor efficiency ε_{th} is defined to be the probability for counting a reaction particle when a 2200 m/s thermal neutron is incident on the foil. It is proportional to the product of the reference thermal cross section σ_{th}, the areal density of the foil, and the solid angle of the particle counters. The flux monitor count rate R_n is then

$$R_n = \varepsilon_{\text{th}} A_{\text{beam}} v_{\text{th}} \int \frac{\phi_n(v)}{v} dv. \qquad (7)$$

Combining Eqs. 4, 5, and 7 we obtain an expression for the neutron lifetime in terms of experimentally measured quantities:

$$\tau_n = \frac{R_n \varepsilon_p L_{\text{det}}}{R_p \varepsilon_{\text{th}} v_{\text{th}}}. \qquad (8)$$

In recent experiments ε_{th} has been the most challenging and systematically limiting of these. Note that the velocity integrals in Eqs. 4 and 7 cancel so to first approximation the neutron spectral flux $\phi_n(v)$ is not needed. It does however appear in important corrections such as self absorption in the foil, which is not perfectly thin as assumed above.

2. Previous Beam Lifetime Experiments

In 1934 Fermi[2] published his famous paper on the theory of beta decay, showing that the weak decay of a neutron into a proton, electron, and antineutrino will occur when energetically possible, *i.e.* when the mass of the parent system is larger than the mass of the daughters. Within a year Chadwick and Goldhaber[3] determined the neutron mass to sufficient precision to show that it exceeds the proton plus electron mass. These two results together predicted free neutron beta decay. It was first observed by Snell and Miller in 1948[4] using a thermal neutron beam. Shortly thereafter Robson, using the beam method, made the first proper measurement of the neutron lifetime at the NRX reactor in Chalk River, Canada.[5] Table 1 summarizes the beam neutron lifetime results through present. Interestingly there has been a steady decline in the mean value as the precision improved.

Table 1. A summary of neutron lifetime measurements using the beam method. When applicable, statistical and systematic errors have been added in quadrature.

Year	location	neutron lifetime (s)	uncertainty (s)	reference
1951	Chalk River, Canada	1110	220	5
1956	Kurchatov Institute, USSR	1040	130	6
1959	Argonne National Lab, USA	1100	160	7
1959	Kurchatov Inst., USSR	1013	26	8
1972	Risö, Denmark	918	14	9
1978	Kurchatov Inst., USSR	881	8	10
1988	Institut Laue-Langevin, France	876	21	11
1988	Kurchatov Inst., USSR	891	9	12
1989	Institut Laue-Langevin, France	878	30	13
1996	Institut Laue-Langevin, France	889.2	4.8	14
2005	NIST, USA	886.3	3.4	1

In 1972 Christensen, et al.[9] achieved a major advance in measuring the neutron lifetime at the Risö reactor in Denmark. Two parallel paddles of plastic scintillator were placed between the poles of a large (60-cm diameter) uniform electromagnet and the neutron beam passed between the paddles. Neutron decay electrons were transported by the magnetic field lines to well-defined regions of the paddles. Backscattered electrons were transported to the opposite paddle so in effect the full electron energy was measured by summing the signals. This approach of detecting a single final state particle with uniform efficiency and a well determined detection region led to a significant improvement in precision.

At the Kurchatov Institute in Moscow, Spivak led a program of beam neutron lifetime measurements that spanned three decades.[6,8,10,12] In these experiments neutron decay recoil protons were accelerated electrostatically to 25 keV and counted with nearly 100% efficiency. The use of proton counting had the important advantage that protons could be stopped prior to detection by applying a small (800 V) potential. This potential had negligible effect on the neutron beam and the backgrounds it produced so a precise background subtraction could be made.

The Sussex-ILL-NIST neutron lifetime program achieved the best precision to date using the beam method.[1,14-16] Here a highly collimated cold neutron beam was passed through the center of of a quasi-Penning trap consisting of 16 nearly identical electrodes in a 4.6 T axial magnetic field (see Figure 2). Three electrodes on each end were held at +800 V, above the maximum proton recoil energy (752 eV), while the central electrodes were at ground. Neutron decay protons were trapped axially by the electrostatic potential and radially by the magnetic field. Periodically (typically every 10 ms) the first three electrodes (the "door") were lowered to ground and a ramp potential applied in the center to flush the trapped protons out through the door. They followed a 9.5° bend in the magnetic field and were accelerated and counted by a silicon surface barrier detector at -30 kV. Downstream of the trap, the neutron beam passed through a 39 $\mu g/cm^2$ layer of ^{6}LiF deposited on a thin (400 μm) silicon foil. The foil was viewed by four silicon surface barrier detectors in a geometry such that, to first order, their combined solid angle was independent of the source spot on the foil. Recoil tritons and alphas from the ^{6}Li$(n,t)^4$He reaction were counted.

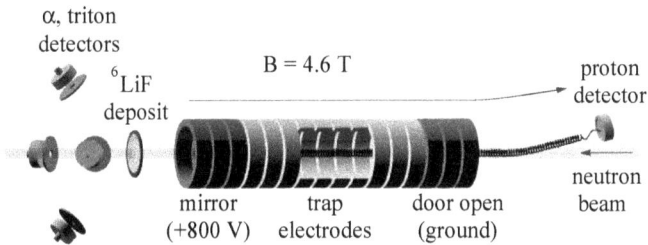

Fig. 2.   Scheme of the Sussex-ILL-NIST beam neutron lifetime experiment.

The proton trap has an inherent difficulty in that the effective length of the trap L_{det} is not well defined. Protons that originate in the central grounded region are fully trapped, but some protons are born at an elevated potential close to the end electrodes and trapped with an efficiency that depends on axial kinetic energy. To address this issue different trap lengths were used, $i.e.$ measurements were made with 3–10 central grounded electrodes. This gives $L_{det} = nl + L_{end}$ where n is the number of central grounded electrodes, l is the length of a single electrode plus spacer, and

L_{end} is the effective length of both end regions, proportional to their physical length and the probability that protons born there will be trapped. To good approximation (due to the axially symmetric design of the apparatus) L_{end}, while not well known, was a constant as the trap length was varied. We rewrite Eq. 8 as:

$$\frac{R_p}{R_n} = \tau_n^{-1} \left(\frac{\varepsilon_p}{\varepsilon_{\text{th}} v_{\text{th}}} \right) (nl + L_{\text{end}}). \tag{9}$$

The ratio R_p/R_n was measured as a function of n and fit to a straight line, obtaining τ_n from the slope without a need to know L_{end}. The experiment completed at NIST in 2003 achieved the smallest uncertainty to date, 3.4 s, for the beam method. The uncertainty was dominated by systematic effects associated with ε_{th}, namely the areal density of the ^{6}Li deposit (0.25%), the ^{6}Li$(n,t)^4$He thermal cross section σ_{th} (0.14%), and the reaction product detector solid angle (0.11%) More information about this experiment can be found in Nico et al.[1]

Details and analysis of the history of neutron lifetime experiments can be found in the recent review by Wietfeldt and Greene.[17]

3. A Comparison of Recent Beam and Bottle Experiments

Figure 3 summarizes recent neutron lifetime experiments using both the beam method and ultracold neutron (UCN) storage bottles. The much discussed dispute among recent UCN measurements[18] was largely resolved in 2012 by downward revisions[19,20] of two previously reported results owing to reanalysis of systematic effects. The UCN results are now in good agreement. Still, the global assessment of all seven experiments is not great, yielding $\tau_n = 880.0$ s with $\chi^2 = 14.0$ for 6 degrees of freedom ($P = 3\%$). A problem remains but it has evolved into a 2.6 σ disagreement between the averages of the beam and bottle results. The field must now address this issue. It is possible that persistent systematic biases have affected either or both of these methods. Clearly, more experiments and systematic studies using both methods are needed to resolve this and push the global uncertainty in the neutron lifetime lower.

4. A Path to a 0.1 s Neutron Lifetime Uncertainty Using the Beam Method

There has been considerable motivation and recent interest in improving the uncertainty on the neutron lifetime to below 0.1 s. Much of the discussion and activity has focused on the UCN bottle method as it currently

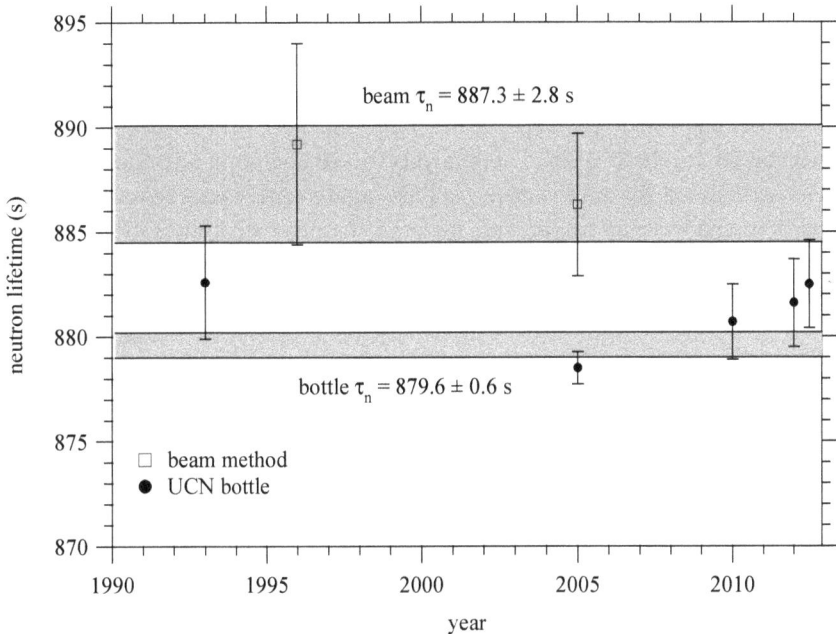

Fig. 3. A summary of recent neutron lifetime measurements using the beam method and ultracold neutron storage bottles, including recent revisions of previously published results.[19,20]

has the best reported precision. However it is important to note that a 0.1 s experiment using the beam method is quite possible. In particular we have in mind a scaled-up version of the Sussex-ILL-NIST apparatus that would be designed and built specifically for such a measurement. This program is quite mature (more than 30 years) and many systematic effects and issues have been thoroughly investigated and are well understood. As a basis for discussion, the relevant portions of the error budget from the NIST experiment[1] are shown in Table 2. The errors separate naturally into those associated with proton trapping/counting and neutron counting by the flux monitor. The limiting sources of error were in the flux monitor thermal neutron efficiency ε_{th}, indicated by asterisks in the table. It has long been realized that a separate absolute calibration of the flux monitor is needed to make a significant improvement on those. A program for such a calibration, using a variety of methods, has been in progress for some time at NIST and collaborating institutions. Within the past year A. Yue and collaborators achieved a successful absolute measurement of ε_{th} with 0.06% relative uncertainty, or 0.5 s in the neutron lifetime, using an α–γ

coincidence technique with a ^{10}B target. They have also developed a plan to improve the method to below 0.01% with a new, larger apparatus (see Yue, *et al.* in these proceedings). With this result in hand it is possible to correct the previously reported neutron lifetime[1] and reduce the overall uncertainty to about 2.3 s. The analysis is now complete and the result has been submitted for publication.[22] This, along with incremental improvements of the existing apparatus, makes a new run with an anticipated 1.0 s uncertainty possible. It is now under development at NIST and data collection is expected to commence in 2014.

A new, larger apparatus will be needed to improve proton counting statistics and systematics to the 0.1 s uncertainty level. Its design and construction are, in principle, straightforward:

- A larger diameter (2×) and longer (2×) segmented proton trap of similar design.
- A larger diameter (5×, *i.e.* 35 mm) neutron beam.
- A correspondingly larger 5 T magnet, with improved uniformity in the trap region.
- A large diameter (52 mm), segmented ion-implanted silicon proton detector to allow the larger neutron beam and eliminate the beam halo correction. A similar detector has been recently demonstrated and will be used in the upcoming Nab experiment at the Spallation Neutron Source.[21]
- Improved electrostatic focusing to reduce the proton backscatter correction.
- Greater neutron capture flux (> 3×), *e.g.* at the new NG-C end position at NIST.

The remaining items in the neutron counting error budget, the flux monitor self absorption and silicon foil neutron scattering, are not eliminated by the absolute calibration of ε_{th} because they depend on the neutron beam spectral flux $\phi_n(v)$. The first step in lowering these is to reduce the thickness of both the deposit and the silicon substrate, at least a factor of two should be possible, and the errors directly scale. Further reduction will come from a more careful and thorough measurement of $\phi_n(v)$, an effort that was treated minimally in the previous NIST experiment as these errors were not limiting then.

A 0.1 s beam neutron lifetime experiment seems feasible using a new proton counting and trapping apparatus that is scaled up but functionally very similar to the existing. On the neutron counting side, an absolute measurement of ε_{th} has been demonstrated at the 0.06% level. The systematic and statistical improvements needed for 0.01% precision have been studied, and preliminary design of a new apparatus for this purpose is in hand. We emphasize that no new technology or untested methodology will be needed for this.

Table 2. Important sources of error from the NIST experiment[1] needing improvement for a 0.1 s beam neutron lifetime experiment. Asterisks indicate those addressed by an absolute calibration of the neutron flux monitor.

neutron counting		proton counting	
Source	error (s)	Source	error (s)
^{6}Li deposit areal density*	2.2	Trap nonlinearity	0.8
^{6}Li cross section*	1.2	Detector backscatter	0.4
Flux monitor detector solid angle*	1.0	Beam halo	1.0
Flux monitor self absorption	0.8	Counting statistics	1.2
silicon foil scattering	0.5		
Counting statistics	0.1		
Quadrature sum	2.9	Quadrature sum	1.9

5. Conclusions

From its origin in 1951 to the present, the beam method for measuring the neutron lifetime has improved in precision from ±220 s to ±3.4 s. Making use of recent achievements in absolute calibration of the neutron flux monitor and improvements to the proton counting apparatus, the NIST program plans further stepwise progress to ±2 s, ±1 s, and eventually ±0.1 s. The latter will require new, scaled-up apparatus for proton counting, neutron counting, and calibrations, but no new technology or methods. Considering the history of neutron lifetime experiments using all methods over the last 60+ years, it is clear that systematic uncertainties have often been inadequately understood and underestimated. Probably, judging from the poor agreement between recent beam and bottle experiments, the problem continues to this day. A reliable value for the neutron lifetime at the 0.1 s precision level will require agreement at that level by the best beam and bottle measurements.

Acknowledgements

The author thanks M. S. Dewey, G. L. Greene, J. S. Nico, W. M. Snow, and A. Yue for very helpful discussions. This work is supported by the National Science Foundation (PHY-0855310, PHY-1205266).

14

References

1. J. S. Nico, *et al.*, Phys. Rev. C **71**, 055502 (2005).
2. E. Fermi, Z. Phys. **88**, 161 (1934).
3. J. Chadwick and M. Goldhaber, Proc. R. Soc. A **151**, 479 (1935).
4. A. H. Snell and L. Miller, Phys. Rev. **74**, 1217 (1948).
5. J. M. Robson, Phys. Rev. **83**, 349 (1951).
6. P. E. Spivak, A. N. Sosnovsky, Y. A. Prokofiev, and V. S. Sokolov, 1956, *Proceedings of the International Conference on the Peaceful Uses of Atomic Energy*, Geneva, 1955 (United Nations, New York), volume 2, p. 33.
7. N. D'Angelo, Phys. Rev. **114**, 285 (1959).
8. A. N. Sosnovsky, P. E. Spivak, Y. A. Prokofiev, I. E. Kutikov, and Y. P. Dobrinin, Nucl. Phys. **10**, 395 (1959).
9. C. J. Christensen, A. Nielsen, A. Bahnsen, W. K. Brown, and B. M. Rustad, Phys. Rev. D **5**, 1628 (1972).
10. L. N. Bondarenko, V. V. Kurguzov, Y. A. Prokofiev, E. V. Rogov, and P. E. Spivak, JETP Letters **28**, 303 (1978).
11. J. M. Last, Arnold, J. Dohner, D. Dubbers, and S. J. Freedman, Phys. Rev. Lett. **60**, 995 (1988).
12. P. E. Spivak, Zhurnal Eksperimentalnoi Teoreticheskoi Fiziki **94**, 1 (1988).
13. R. P. Kossakowski, R P. Grivot, P. Liaud, K. Schreckenbach, and G. Azuelos, Nuclear Physics A **503**, 473 (1989).
14. J. .P Byrne, P. G. Dawber, C. Habeck, S. Smidt, J. Spain, and A. Williams, Europhys. Lett. **33**, 187 (1996).
15. J. Byrne, *et al.*, Phys. Rev. Lett. **65**, 289 (1990).
16. M. S. Dewey, *et al.*, Phys. Rev. Lett. **91**, 152302 (2003).
17. F. E. Wietfeldt and G. L. Greene, Rev. Mod. Phys., **83**, 1173 (2011).
18. K. Nakamura, *et al.* (Particle Data Group), J. Phys. **G 37**, 075021 (2010).
19. S. S. Arzumanov, L. N. Bondarenko, V. I. Morozov, Yu. N. Panin, and S. M. Chernyavsky, JETP Lett. **95**, 224 (2012).
20. A. Steyerl, J. M. Pendlebury, C. Kaufman, S. S. Malik, and A. M. Desai, Phys. Rev. C **85**, 065503 (2012).
21. S. Wilburn, unpublished report (2011).
22. Yue, A. T., et al., http://arxiv.org/abs/1309.2623v1 (submitted to Phys. Rev. Lett.).

Overview of Magnetic Trapping
Neutron Lifetime Experiments

P. R. HUFFMAN

Physics Department, North Carolina State University,
Raleigh, NC 27699, USA
paul_huffman@ncsu.edu

Experiments utilizing magnetic trapping techniques to measure the neutron lifetime are becoming increasingly popular. These techniques have a history dating back to the mid 1950's. I will provide a brief history of the use of magnetic fields for manipulation and storage of neutrons as it relates primarily to present-day lifetime measurements.

Keywords: Neutron lifetime; magnetic trapping

1. Introduction

The feasibility of using magnetic fields to manipulate neutrons was first presented by W. Paul at a conference in 1951.[1] Here, he discussed using a series of three magnetic poles to focus low-energy neutrons within a given solid angle using the $\vec{\mu}_n \cdot \vec{B}$ interaction; neutrons with spins parallel to a magnetic field experience a force pushing them away. Here, $\mu_n = -0.96623647 \times 10^{-26}$ J/T [4] is the magnetic moment of the neutron which is antiparallel to the neutron spin, and $\vec{B}$ is the magnetic field. A neutron in a 1.66 T magnetic field experiences a $\vec{\mu}_n \cdot \vec{B}$ interaction energy of 100 neV. This is comparable to the Fermi potential energy of neutrons interacting with a material surface, typically < 300 neV,[2] as well as the gravitational potential energy corresponding to 1.0 m in height, 103 neV.

About ten years later, Vladimirskiĭ first proposed a field geometry where one could obtain spacial confinement of neutrons using magnetic fields.[3] He made estimates on the number of neutrons one could confine in such a magnetic bottle and considered the adibacity conditions necessary for maintaining the orientation of the neutron spin relative to the field. A neutron will precess around the magnetic field lines with a Larmor frequency $|\gamma|B$, where $\gamma = 1.832 \times 10^8$ s^{-1}T^{-1} is the gyromagnetic ratio.[4] As the neutron moves through field gradients, the spin will follow the field provided that the Larmor frequency is large compared to the rate the field changes, $|\gamma|B \gg \frac{dB}{dT}/B$.

The feasibility of storing neutrons using magnetic fields was first experimentally tested in the mid 1970's by Kosvintsev et al. using a 10 cm diameter cylindrical copper tube 2.4 m long.[5] The tube was loaded with ultracold neutrons (UCN) from the SM-2 reactor and they observed storage lifetimes of order 25 s in the tube itself for UCN with energies from 0 neV to 30 neV. The copper end cap was then replaced with neutron absorbing polyethylene, thereby significantly reducing the lifetime of neutrons in the bottle. An electromagnet was placed at the same end as the polyethylene and energized along with a holding field coil around the copper tube to maintain the spin direction. They were able to show that data taken with the electromagnet on was consistent with data taken with the copper end cap as opposed to the polyethylene, thereby confirming that magnetic fields can be used to store low-energy neutrons.

This same group then built and tested what I would consider to be the first gravito-magnetic trap.[6] This cylindrical bucket-shaped trap, 80 cm in diameter, 80 cm tall, used magnetic coils along the bottom and walls of the cylinder to produce field gradients in order to contain the neutrons. The gravitational potential provided confinement in the vertical direction. Ultracold neutrons with energies between 0 neV and 9 neV were loaded, and later detected, through a removable "stopper" in the lower central region of the trap. Using this geometry, they were able to store (1.05 ± 0.15) neutrons in the trap with a storage lifetime of (25 ± 10) s. This was the first true storage of neutrons in a gravito-magnetic trap.

In an attempt to develop a trap capable of storing neutrons for times approaching the neutron lifetime $(\tau_n = 880 \text{ s})$,[7] Abov et al. built a significantly larger trap, 1.04 m in diameter, 0.34 m tall.[8] Although similar in design to the trap of Kosvintsev et al.,[6] they paid particular attention to the field design so that it would have no zero-field regions; it was initially believed that the trap storage time of Kosvintsev et al. was limited by losses from spin-flips in these zero-field regions. In these designs, the magnetic field falls off roughly exponentially as the neutron moves away from the wall. To decrease spin-flip losses, an additional electromagnet was added near the top of the trap. With this trap, they were able to confine on average 3.6 neutrons in the trap with a storage lifetime of (303 ± 37) s, roughly an order of magnitude improvement from the previous work.

Another major advance was made by this same group when they real-

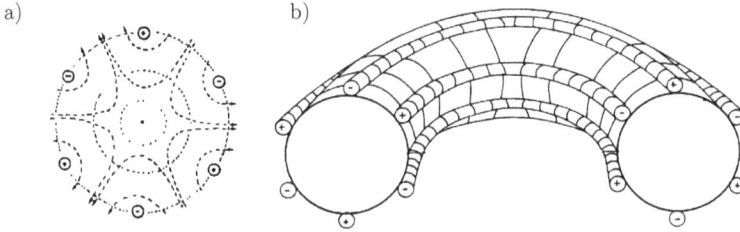

Fig. 1. Sextupole magnetic trap configuration. a) Cross section and field of an ideal hexapole. b) Sextupole torus geometry.[12]

ized that spin-flip losses could not fully account for the short lifetimes in their previous measurements. They speculated that the primary loss mechanism was actually due to neutrons whose "energy range exceeds the magnetic storage energy limit".[9] These quasi, or marginally trapped neutrons can escape the trap undetected during a measurement cycle, thus giving a short storage lifetime. By placing a neutron absorber at the top of their trap to absorb these higher energy neutrons, they were able to show that the storage lifetime had two distinct components, one with a time constant of about 200 s, and a longer-lived one with a lifetime greater than 700 s. A later reanalysis of this data provided a range for the neutron storage time of (830^{+430}_{-205}) s, with an initial number of the long-lived neutrons having a mean value of 0.3 neutrons per run.[10] These experiments laid the foundation for several present-day neutron lifetime measurements that I will discuss below.

2. Three-Dimensional Magnetic Confinement

The above-mentioned traps all rely on gravity for confinement in the vertical dimension. In 1963, Heer described a geometry for a different type of "magnetic bottle" that consisted of six straight wires with alternating currents symmetrically arranged in a hexagonal geometry as shown in Fig. 1.[11] These wires are formed into a toroidal or doughnut shaped geometry. While his paper is aimed primarily at neutral atom confinement, he concludes by noting that neutrons could be stored in such a ring as well, although thermal moderation to the low energies required had not yet been obtained.

Such a geometry was realized for neutrons about 15 years later using a variation of this hexapole geometry. Kügler et al. demonstrated the feasibility of storing free neutrons using a sextupole torus geometry with the addition of decapole field compensating coils.[13,14] They were able to store approximately 1000 neutrons per run with longitudinal velocities between 8 m/s and 15 m/s and radial velocities up to 4 m/s for times up to 45 minutes. Initial measurements were limited by losses through betatron oscillations; longitudinal momentum is transferred to the radial direction and neutrons are no longer in stable orbits. These oscillations were minimized by collimation of the neutrons as well as adding additional field gradients. The first competitive measurement of the neutron lifetime using magnetic trapping techniques was performed using this geometry. In the end, they were able to report a measurement of the lifetime, $\tau_n = (876.7 \pm 10)$ s,[12,15] that was competitive with both beam and bottle type experiments at the time.

Heer[11] and others[16,17] go on to discuss several additional trap geometries of potential use for confinement of neutrons. The simplest of these is two circular current loops arranged in an anti-Helmholtz configuration as shown in Fig. 2. Although high field gradients can be obtained using this type of geometry, a zero-field region always exists in the center where one could lose neutrons. Other geometries include a spherical hexapole, an Ioffe configuration, and a "baseball" trap with a current loop in a shape similar to a seam on a baseball (not pictured).

While the author is not aware of anyone using either the anti-Helmholtz or baseball trap geometries for three-dimensional confinement of neutrons, both the spherical hexapole and the Ioffe configurations have been used. As part of the thesis work of N. Niehues in the early 1980's, a group in Bonn attempted to load neutrons into a spherical hexapole trap using the superthermal process in helium.[18] The trap had a 10 cm diameter trapping region with a maximum field of 2.5 T. They were ultimately unsuccessful in demonstrating trapping using this geometry. The primary reason was attributed to phonon upscattering in the 1.2 K helium bath.

The Ioffe trap design shown in Fig. 2 is the simplest type of multipole trap and consists of four current bars that create a quadrupole field to provide radial confinement, combined with two solenoids with the same current sense to provide longitudinal confinement. As in the sextupole geometry discussed above, one can in principle increase the phase space available to neutrons in the trap by going to a higher order multiple as shown in Fig. 3. For a fixed field strength and radius, one can gain up to a factor of two in the number of trapped neutrons if one goes to a significantly high-order multipole.

One must also, however, take the experimental geometry into account

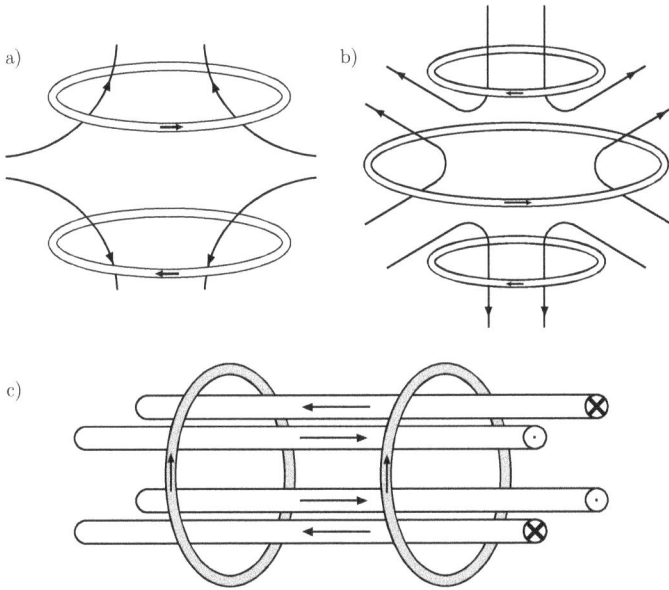

Fig. 2. Magnetic trap geometries: a) anti-Helmholtz, b) spherical hexapole, and c) Ioffe.

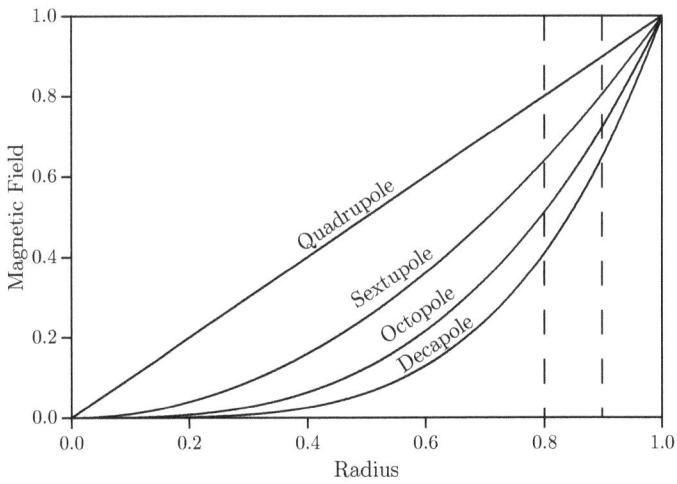

Fig. 3. The radial dependence of selected multipole fields. The vertical dashed lines are for reference as described in the text.

when determining the appropriate multipole. In Fig. 3, there are two vertical dashed lines drawn for reference that could represent the actual radius of a trapping region. This smaller radius could arise for example due to the presence of a vacuum can separating the liquid helium surrounding the magnetic coils from the experimental region. At smaller radii, the field falls off faster for higher order multipoles; at a radius of 80 %, the trap depth of a decapole is roughly 50 % smaller than a quadrupole. Since the number of neutrons trapped typically scales faster than the magnetic field (i.e. the number trapped in a quadrupole field scales as $B^{3/2}$), the quadrupole geometry would be preferable in this case where the field at the edge of the trapping region is significantly higher than for higher multipole fields.

An Ioffe configuration has been used to trap neutrons produced using the superthermal technique. In this experiment, a beam of 0.89 nm neutrons is incident on a superfluid ^{4}He target within the minimum field region of an Ioffe-type magnetic trap. Some of the neutrons are downscattered in the helium to energies < 200 neV, and those in the appropriate spin state become trapped. The inverse process is suppressed by the low phonon density in < 200 mK helium, allowing the neutron to travel undisturbed. When the neutron decays however, the energetic electron ionizes the helium, producing scintillation light that is detected using photomultiplier tubes. This group successfully demonstrated the trapping of ultracold neutrons (UCN) at the National Institute of Standards and Technology Center for Neutron Research (NCNR)[19,20] and have since upgraded the apparatus in preparation for a neutron lifetime measurement. Additional information on this experiment can be found in these proceedings.

3. Gravito-Magnetic Traps

As discussed earlier, since the gravitational and magnetic energy scales are comparable, one can use a combination of these interactions to confine neutrons.[21,22,24,26] There are four such experiments either underway or being constructed that utilize gravito-magnetic traps and each was discussed in detail at this workshop. In general, these traps use high multipole field geometries to produce large field gradients that provide confinement at the walls of the trap, with the vertical confinement provided by the gravitational interaction. Each typically has a uniform holding field to preserve the spin direction of the neutrons.

An experiment based at the Institut Laue-Langevin (ILL) uses an array of permanent magnets in a cylindrical geometry to produce a 1 T depth trap.[21] UCN are loaded into the trap originally through a valve along the cylindrical axis at the bottom of the trap, and now from the top by lowering a material bottle filled with UCN into the volume. The walls of the trap are coated with Fomblin oil to retain any spin-flipped neutrons, which are detected as they leave the trap through the bottom hole. They were able to measure storage times comparable to that of the neutron lifetime. Presently, they are in the process of analyzing a set of lifetime data taken in 2007 as well as constructing a new trap that is approximately 15 times larger in volume. The statistical accuracy of the 2007 lifetime data is less than 2 s, while the larger trap aims for less than 1 s.

A second experiment that uses permanent magnets is being constructed and tested at Los Alamos National Laboratory at their UCN source facility.[22,23] Here, a large Halbach array of permanent magnets arranged in a non-symmetric geometry provides confinement, with gravity providing vertical confinement. The trap is roughly bathtub shaped and has a volume of 0.6 m^3 and field strength of 1 T at the surface of the trap. UCN will be loaded into the trap from the bottom using a trap-door approach. The trap is presently under construction, with an initial proof-of-principle run having been recently performed. Additional details can be found in these proceedings.

The HOPE experiment based at the ILL uses a design similar to that of the Ioffe configuration, but with a permanent Halbach octupole configuration instead of the quadrupole coils.[24,25] Their trap, presently under construction, will have a trap depth of 1.3 T with an 8 l volume. The trap will be mounted vertically with UCN loaded through a magnetic valve at the bottom of the trap. Additional details can also be found in these proceedings.

The fourth experiment is known as Penelope and employs alternating rings of superconducting magnets to produce large field gradients at the surface of the trap.[26,27] Two concentric sets of these rings are stacked vertically producing a trap with a depth of 2 T, with no zero-field regions. The decay protons are guided to a scintillator at the top of the trap and detected. Initial testing of a magnet coil prototype has been performed and they are expecting to begin commissioning the magnet in the coming year. Again, see their contribution for additional information.

4. Conclusions and Outlook

Several unique approaches for measuring the lifetime of the neutron using magnetic traps are presented in these proceedings. As these experiments evolve, it is clear that one must fully understand the systematic effects associated with magnetic traps and avoid the situation that the material bottle lifetime experiments are presently in where one has several measurements, but clearly the systematics are not under control.

While the systematics vary between different magnetic trapping experiments, the single common systematic effect that everyone must understand is related to the dynamics of neutrons in the trap, and especially the behavior of the quasi or marginally trapped neutrons. Several groups have performed detailed analyses of neutron trajectories in magnetic traps[22,28–31] and have highlighted the difficulties in performing these simulations. Understanding these dynamics is a complicated problem and one must understand the orbits as well as the efficiency of any cleaning processes in order to have confidence in the measurement results.

Magnetic trapping experiments do however have distinctly different systematic effects than those of either the beam or bottle techniques. In addition, these experiments offer the possibility of reaching higher precisions as well. Having such independent experiments will be essential for both resolving the discrepancies in the present-day lifetime measurements as well as providing confidence in these measurements.

References

1. W. Paul, Int. Conf. on Nucl. Phys. and the Phys. of Fund. Part. (1951).
2. Golub, Richardson, and Lamoreaux, *Ultra-Cold Neutrons*, Taylor & Francis (1991).
3. Vladimirskiĭ, Soviet Physics JETP **12(4)**, 740 (1961).
4. NIST Standard Reference Database 121, 2010 CODATA recommended values.
5. Kosvintsev *et al.*, JETP Letters **23(3)**, 118, (1976).
6. Kosvintsev *et al.*, JETP Letters **27(1)**, 65, (1978).
7. Beringer *et al.* (Particle Data Group), Phys. Rev. D **86**, 010001 (2012).
8. Abov *et al.*, Yad. Fiz. **38**, 122, (1983).
9. Abov *et al.*, JETP Letters **44(8)**, 472 (1986).
10. Yu. G. Abov, V. V. Vasiliev, and O. V. Schvedov, Physics of Atomic Nuclei, **63**, 1305 (1990).
11. C. V. Heer, Rev. Sci. Inst., **34**, 532 (1963).
12. F. Anton *et al.*, Nucl. Instrum. Methods Phys. Res. A **284**, 101 (1989).
13. K.-J. Kügler, W. Paul, and U. Trinks, Phys. Lett. B **72**, 422 (1978).
14. K.-J. Kügler, K. Moritz, W. Paul, and U. Trinks, Nucl. Instrum. Methods Phys. Res. A **228**, 240 (1985).
15. W. Paul *et al.*, Z. Phys. C **45**, 25 (1989)
16. I. M. Matora, Sov. J. Nucl. Phys. **16**, 349 (1973).
17. T. Bergeman *et al.*, Phys. Rev. A **35**, 1535 (1987).
18. Niehues, N. *Untersuchungen an einer magnetischen Flasche zur Speicherung*

von Neutronen. (Investigations on a magnetic bottle for the storage of neutrons.) Thesis, Friedrich Wilhelm Univ. Bonn (1983).
19. P. R. Huffman *et al.*, Nature **403**, 62-64 (2000).
20. C. R. Brome *et al.*, Phys. Rev. C **63**, 055502 (2001).
21. V. Ezhov *et al.*, Nucl. Instrum. Methods Phys. Res. A **611**, 167 (2009).
22. P. L. Walstrom *et al.*, Nucl. Instrum. Methods Phys. Res. A **599**, 82 (2009).
23. A. Saunders *et al.*, these proceedings.
24. K. K. H. Leung and O. Zimmer, Nucl. Instrum. Methods Phys. Res. A **611**, 181 (2009).
25. K. Leung *et al.*, these proceedings.
26. S. Materne *et al.*, Nucl. Instrum. Methods Phys. Res. A **611**, 176 (2009).
27. R. Picker *et al.*, these proceedings.
28. K. J. Coakley, Nucl. Instrum. Methods Phys. Res. A **406**, 451 (1998).
29. K. J. Coakley *et al.*, J. Res. Natl. Inst. Stand. Techol. **110**, 367 (2005).
30. J. D. Bowman and S. I. Penttila, J. Res. Natl. Inst. Stand. Technol. **110**, 361 (2005).
31. A. Steyerl *et al.*, Phys. Rev. C **86**, 065501 (2012).

Some Thoughts Concerning the Statistics and Ultracold Neutron Source Requirements for the UCNτ Experiment

A. R. YOUNG

Physics Department, North Carolina State University,
Raleigh, NC 27695, USA
aryoung@ncsu.edu

Some general considerations concerning the necessary number of measured decays to reach a target precision of roughly 10^{-3} for a UCN storage neutron lifetime experiment are presented. In the context of the UCNτ experiment, a brief summary of UCN sources available at present and those expected to be operational are also tabulated, together with a "lumped-element" analysis of the target transport properties of the guide system used to load the experiment.

Keywords: Neutron; ultracold; lifetime; source

1. Introduction

The purpose of this presentation is to outline some general considerations for the statistics required for the UCNτ experiment to provide a roughly 1 s measurement of the neutron lifetime. We review roughly what statistics and UCN densities might be required, what sources are currently and will be available to load this experiment, and what source and guide parameters have the greatest impact on achieving the statistics goals.

The number of detected decays required to achieve a target precision of 10^{-3} in a neutron lifetime experiment depends on the details of the experimental method, but we can typically expect that Poisson statistics will hold and that the total number of **detected** neutrons must be of order 10^6 or more to achieve this uncertainty. To be concrete, we select the case of a UCN storage experiment in which the contents of the trap are measured after storage times t_1 or $t_2 = t_1 + 2.2\tau$, where the neutron lifetime is τ. When the number of runs performed with each storage time are the same, the additional delay between t_1 and t_2 of 2.2τ produces the minimum statistical uncertainty, a detail exploited in recent experiments.[1] In this case, at least 2.1×10^6 detected decays are required for a precision of one part in a thousand in the lifetime.

This estimate must be tempered by the anticipation of one or more sources of systematic error which must be adequately characterized. Examples of some of these potential sources of error are marginally trapped neutrons, depolarization, and spectral evolution of the UCN coupling to their storage and detection efficiency.[1-5] Ideally, one has the opportunity to characterize these sources of error with additional, dedicated measurements (which must have superior statistical precision if they are not to expand the uncertainty significantly). As an arbitrary assessment of the additional requirements, we assume that adequate characterization might require roughly an order of magnitude more counts than that required for our targeted statistical precision and would be analogous to the characterization of wall loss effects in past UCN storage lifetime experiments.[1,3,6]

Although for many systematic errors, an assessment is possible without resorting to studies requiring a series of extra lifetime measurements (for example, the UCN spectrum might be directly measured at different times to assess the impact of spectral evolution), we assume here that one or two effects will emerge which require more detailed assessment. If we also assume that the data obtained during these dedicated runs are not "reused" to improve the statistical uncertainty of a final result, then to achieve 10^{-3} precision with one such systematic error being characterized, about 2.8×10^7 detected decays are required, and if two sources of error must be characterized, this number increases to about 6.4×10^7 detected decays. As a relatively conservative starting point, we therefore adopt 6×10^7 detected decays as a target for the statistical sample necessary for a 0.1% measurement of the neutron lifetime.

Focusing on UCNτ, recent UCN storage lifetime experiments typically have a run cycle which incorporates a (1) loading, (2) cleaning, (3) storage, (4) stored UCN measurement, and (5) background measurement phase. The time required for each of these phases varies from experiment to experiment, but for UCNτ we anticipate a loading time to "saturate" the trap of 200 to 300 s, cleaning times less than about 200 s, measurement times less than 200 to 300 s (much faster if the V absorption method is used), and background measurement times of several hundred seconds. If we select "optimal" choices for two storage times of 0 and 1940 s, reasonable run cycles might be then be 1000 s or 2940 s (average about 2000 s). Assuming that one runs for 30 days, with a 50% duty cycle, this means that one has available about 648 runs, and the average number of detected UCN per run must be 9.3×10^4. Assuming the UCN detection scheme has an

Table 1. Details and references for each entry are in the text. The density quoted for the sources are for unpolarized UCN except the TRIUMF source, which is polarized.

Source	Available	Density in storage vol UCN/cm^3	Useful Current 10^4 UCN/s	Source Storage Time (s)
Available now:				
LANL Area B	present	52	$\simeq 10$	40
ILL turbine	present	>39	$\simeq 100$	few sec
ILL LHe	present	>55	4	67
RCNP	present	26	3.2	81
PSI	present	23	4.2	90
			(20M/480s)	
Mainz	present	25	0.42	few sec
			(0.35M/300s)	
Planned:				
PULSTAR	2014+	>30	>10	few sec
FRM II	2015+	5000	3000	Probably secs
TRIUMF	2016	1500*	100*	150
WWR-M	2016	12000	7000	10

*polarized UCN

efficiency of at least 0.25 (characteristic of some of the documented guide and detector systems utilized by the LANL group[7]), this means the trap must be loaded with roughly 8.4×10^5 storable UCN before cleaning. Noting that the trap volume is 7×10^5 cm^3, this permits us to identify a conservative target density in the storage trap for lifetime studies at the 10^{-3} level: $\simeq 1.3$ UCN/cm^3, in an energy range between 0 and 50 neV in the lifetime storage volume, if the trap is unloaded to a conventional UCN detector. For the in *situ* detection scheme proposed for the UCNτ experiment using a vanadium absorber, the detection efficiency could be much higher and the target density could approach 0.33 UCN/cm^3.

2. UCN Sources

We next attempt a cursory review of currently available and planned UCN sources, worldwide. The challenge is to produce densities of 1.3 UCN/cm^3 in the very low energy range from 0 to 50 neV, in a volume of roughly 7×10^5 cm^3. The large volume implies that both UCN integrated flux or "current" and limiting density will be important properties of the source. The source properties discussed in this section are also summarized in Table 1, and we divide the sources into two categories: those that are available now, and those that are planned. We note that most sources load experiments through 7 to 8 cm diameter guides. We note the exceptions to this below.

Available now:

- The LANL Area B source:[8] This spallation source couples roughly 1.5 liters of solid deuterium to the 800 MeV proton beam at LAN-SCE, and produces up to $\simeq$ 52 UCN/cm^3 (the 2010 run cycle) in the 80 liter storage volume coupled to the source (with stainless steel Fermi potential). This 80 l volume has a characteristic storage time of about 40 s. The useful UCN current one can extract from the source is about 10^5 UCN/s. The solid deuterium in the source is shuttered off from the storage volume between beam bursts, to minimize losses.

- The ILL turbine source:[9] This source utilizes VCN extracted from a liquid deuterium source coupled to the 58.3 MW research reactor at the Institut Laue-Langevin. The VCN are "downshifted" in the Steyerl turbine, producing a limiting detected density of $\simeq$ 40 UCN/cm^3 in a 2.8 liter stainless steel volume coupled directly to the output of the turbine. This source still has the largest continuous UCN current (1-2 $\times$ 10^6 UCN/s in roughly 7 cm diameter guides) at present. The UCN lifetime in the source system is very short (a few seconds or less) in the typical operating conditions in which the limiting density is achieved.

- The ILL LHe source:[10] This source utilizes the H172a cold neutron beamline at the Institut Laue-Langevin to produce UCN in superfluid He at roughly 0.7 K. A limiting density of at least 55 UCN/cm^3 develops in the source storage volume of 5 liters, with a storage lifetime of 67 s. This results in an extracted (through a 2.4 cm guide) UCN population in the trap of 2.5 $\times$ 10^5 UCN and a time-averaged, extracted UCN current of about 4 $\times$ 10^4 UCN/s possible as of November, 2011. An additional factor of 2 improvement in the measured UCN density has since been achieved in an upgraded version of their source.

- The PSI source:[11] This spallation source couples 30 liters of solid deuterium to the 600 MeV proton beam at PSI, and currently produces up to 23 UCN/cm^3 in an external, 25 liter, stainless steel volume coupled to the source. The spallation cycle involves producing 1 pulse every 480 s, loading a source storage volume of 2×10^6 cm^3 with a storage lifetime of 90 s. The continuous current one can extract is up to about 4.2×10^4 UCN/s, however, to fill a lifetime experiment, one would probably optimize the filling procedure to minimize the impact of losses in the storage and guide system.

- The RCNP source:[12] This spallation source couples 8 liters of superfluid He to the 400 MeV (at 1 uA) proton beam at RCNP and produces a density of 26 UCN/cm^3 in the 120 liter volume connected to the source. The source operates essentially continuously, developing the limiting density over a characteristic storage lifetime of 81 s. The UCN current produced by the source is roughly 3.2×10^4 UCN/s.

- The Mainz source:[13] This source is coupled to the pulsed TRIGA Mark II reactor at Mainz, with a steady state power of 100 kW or in a pulse mode (up to 12 pulses per hour) with a maximum power of 250 MW. When pulsed, the system is capable of filling a 1.7 liter storage volume coupled directly to the source to 25 UCN/cm^3. The team projects up to 3.7×10^5 UCN/cm^3 produced per pulse, but with a relatively short storage time in the system (less than about 10 s in the 20.5 liter volume of source and guides), because the UCN are not isolated from the solid deuterium between pulses (at present). This leads to a maximum, time average UCN current of 1.3×10^3 UCN/s, but to fill a lifetime experiment, one would clearly optimize the filling time to minimize the impact of losses.

Planned:

- The PULSTAR source:[14] This source couples 1 liter of solid deuterium to the 1 MW PULSTAR reactor at North Carolina State University. It will run continuously, with expected initial useful densities in an external trap in excess of 30 UCN/cm^3 and useful UCN currents in excess of 1×10^5 UCN/s. Construction is essentially complete, with the source in the process of safety approval and could be operational as early as 2014.

- The FRM II source:[15] This source situates 250 cm^3 of solid deuterium in a through-guide inside the biological shield of the 20 MW FRM II research reactor at the Technical University of Munich. The source will run continuously, with a projected density at the exit of the guide system of 5000 UCN/cm^3 and a current of about 6×10^7 UCN/s. This source is under review before final installation can occur and has a goal of being operational in 2015.

- The TRIUMF source:[16] The TRIUMF source is an upgrade of the RCNP source coupled to the 500 MeV proton beam at TRIUMF. The projected storage lifetime is 150 s for the upgrade, with increased proton beam current, higher guide potentials, and improved geometry, increasing the density in the source storage volume to 1300 polarized UCN/cm^3 and producing a UCN current of 10^6 polarized UCN/s. This source is expected to be operational in 2016.
- The WWR-M source:[17] This source will be located inside the thermal column of the 18 MW WWR-M reactor in Gatchina and features a 35 l superfluid He converter at roughly 1.2 K, with an effective UCN storage lifetime in the source of 10 s. This source will produce UCN continuously, with an expected density of 12000 UCN/cm^3 in a 35 l bottle connected to the source and a useful current from the source of about 8×10^7 UCN/s. The source is expected to be operational in 2016.

3. Comments Concerning Loading a UCN Storage Lifetime Experiment

The above list suggests that there are already a number of sources which might achieve the target density of 1.3 UCN/cm^3 in UCNτ and be suitable for other UCN lifetime experiments. In order to understand the constraints and to compare sources, however, some analytical framework to evaluate how source and guide parameters affect the density produced in a storage experiment is helpful. Intuitively, there are a few points which we expect to be relevant. The first is that, for large volume storage experiments, not only the limiting density, but the UCN current (integrated flux) will be important. For example, in almost every case for available sources (except perhaps the ILL turbine source), the density achieved in the UCNτ volume in a 200s loading period is actually determined by the available UCN current, not the limiting density of the source. The second is that the accumulation rate in the source-and-experiment system is determined in part by the operational UCN storage time in the source. The point here is that, even though the UCN storage experiment should have a very long storage time (ideally comparable to the neutron lifetime), the guide system which loads the experiment may not have such a long lifetime. If this UCN survival time in the guides is much less than the operational UCN lifetime in the source, the delivered UCN to the experiment will be significantly reduced. Therefore, the storage lifetime in the source provides a rough standard for guide performance, with longer required source storage times creating higher demands on the guide delivery system.

In what follows, we will concentrate on the case of UCNτ at the LANL Area B source, but some of our results are reasonably generic. In order to analyze the dependence on loading guide parameters, we will provide a "lumped-element" analysis of the source, coupling guides and lifetime experiment storage volume. This approach does not supplant the need for detailed Monte Carlo, and is primarily helpful for indicating the general dependence of the final density in the lifetime experiment on source and guide parameters.

The basic components of the system are depicted in Fig. 1. These components can be modeled by the following set of equations:

$$\frac{dn_1}{dt} = -(C + V_1/\tau_1)n_1 + Cn_2 + I \tag{1}$$

$$\frac{dn_2}{dt} = Cn_1 - ((1+\alpha)C + V_2/\tau_2)\, n_2 + \alpha Cn_3 \tag{2}$$

$$\frac{dn_3}{dt} = \alpha Cn_2 - (\alpha C + V_3/\tau_3)\, n_3 \tag{3}$$

where n_i, V_i and τ_i are the density, volume and UCN survival time for region i of the experiment, C is a UCN conductance (cm^3/s) for standard guides (6.9 cm diameter), I is the useful UCN production rate and α is a loading efficiency for a vertical section of guide into the trap.

To define our production rate for this model, we adjust the parameter I so that, for the measured loss rates in the source storage volume, we reproduce the typical LANL source density when the source system is sealed off from the external guide system, establishing $I \approx 1 \times 10^5$ UCN/s. We estimate the conductance through the horizontal guide based on the conductance through an aperture (since our guides presumably have very high specularity, and after a few 10's of cm of straight guide, do not have large changes in their angular distribution or mean forward velocity). The UCN in the horizontal guide section have a spectrum with a stainless steel cutoff energy (≈ 180 neV). If one uses an estimate $C = A\bar{v}/4$, with a $v^2 dv$ distribution giving $\bar{v} = 440$ cm/s, then $C = 2300$ cm^3/s for 3" guide. Although these are very crude approximations, similar assumptions permit reasonable modeling of the time dependence of the density in the UCNA experiment.

To track the useful density which can be loaded into the trap, we make two (approximate) adjustments: we scale the useful production rate to account for the (reduced) fraction of UCN which can be loaded into the trap

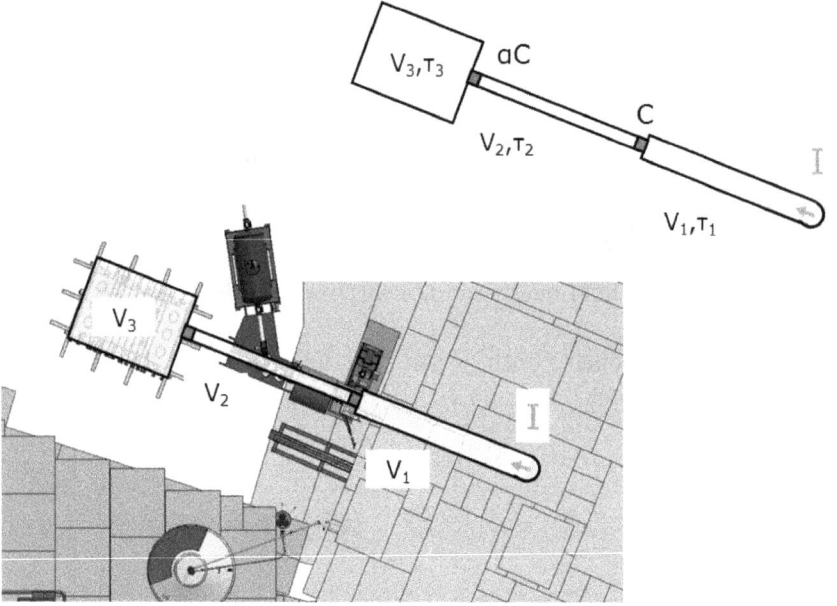

Fig. 1. Model Geometry for the Area B source coupled to the UCNτ experiment (Color Plate 1).

with energies between 0 and 50 neV, and we assume a loading (transmission) efficiency for the trap of α to account for reduced conductance into the magnetic trap relative to the horizontal guide sections.

The production scaling arises in part because only one spin state can be magnetically trapped, resulting in a (crude) factor of two reduction in the useful production rate, and in part from the reduced fraction of UCN which can enter the trap (we assume that the maximum energy of the UCN is reduced to 50 neV by gravity at the entrance to the trap, and is therefore storable). Hence, we have a "trappable fraction" of

$$f = 0.5 \times \left[\frac{E_{ss}^{1.5} - (E_{ss} - 50)^{1.5}}{E_{ss}^{1.5}} \right] \tag{4}$$

which yields a value of $f = 0.19$ with $E_{ss} = 180$ neV and the trapping potential 50 neV. Our effective production rate then becomes $I_e = fI = 1.9 \times 10^4$/s for the Area B source at LANL. We also incorporate a factor of $\alpha = 0.5\epsilon$, where the factor of 0.5 accounts for the slowed UCN population having a reduced conductance at the entrance of the trap, and where ϵ is a transport factor associated with the loading geometry.

The equilibrium solutions are determined assuming the rates of change

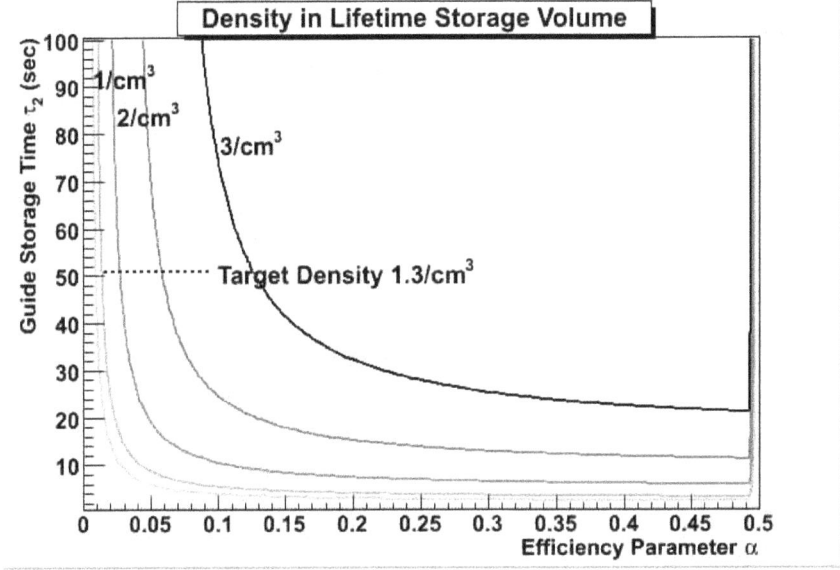

Fig. 2. Model results for the LANL Area B source, where we use approximate values $I_e = 1.7 \times 10^4/\text{s}$, $C = 2300 \text{ cm}^3/\text{s}$, $\tau_1 = 30\text{s}$ and $\tau_2 = 880\text{s}$ (Color Plate 1).

evolve to zero, hence these assumptions are not appropriate for sources operated a "pulsed mode", where one must determine an optimal loading time to trade off between guide and source storage losses and the loading time of the storage experiment. The solutions are

$$n_1 = \frac{I_e}{\Gamma_1} + \frac{C}{\Gamma_1} \left[\frac{\alpha C + V_3/\tau_3}{\alpha C} \right] \left[\frac{1}{1 - \frac{\alpha^2 C^2}{\Gamma_2 \Gamma_3}} \right] \frac{\alpha C^2 I_e}{\Gamma_1 \Gamma_2 \Gamma_3}, \tag{5}$$

$$n_2 = \left[\frac{\alpha C + V_3/\tau_3}{\alpha C} \right] \left[\frac{1}{1 - \frac{\alpha^2 C^2}{\Gamma_2 \Gamma_3}} \right] \frac{\alpha C^2 I_e}{\Gamma_1 \Gamma_2 \Gamma_3}, \tag{6}$$

$$n_3 = \left[\frac{1}{1 - \frac{\alpha^2 C^2}{\Gamma_2 \Gamma_3}} \right] \frac{\alpha C^2 I_e}{\Gamma_1 \Gamma_2 \Gamma_3}, \tag{7}$$

with

$$\Gamma_1 = C + V_1/\tau_1, \tag{8}$$

$$\Gamma_2 = (1 + \alpha)C + V_2/\tau_2, \tag{9}$$

$$\Gamma_3 = \left[1 - \frac{C^2}{\Gamma_1 \Gamma_2} \right] (\alpha C + V_3/\tau_3). \tag{10}$$

The solutions depicted in Eqs. 5 - 7 help us visualize the impact of the performance of the loading geometry. Specifically, we note that for reasonable parameters, it is roughly true that $n_3 \propto \alpha I_e$, so the loaded density just scales with the transport efficiency into the trap and the production rate. As Fig. 2 indicates, the solution also depends strongly on the lifetime in the transport region, with significant losses when τ_2 falls much below the lifetime in the source $\tau_1 = 30$ s. In particular, in this model as τ_2 falls below about 10 s, the required loading efficiency to achieve our benchmark of 1.3 UCN/cm^3 increases very rapidly. Although it is encouraging that even for modest performance in the delivery guide system ($\tau_2 = 10$s and α=0.05) the benchmark density can be achieved, in this very crude model the numerical values for the density probably can be treated only as an order of magnitude estimate. As details of the transport system are fixed, a more accurate and predictive model can be produced.

4. Conclusions

In this Proceedings, we've reviewed the general needs for a UCN storage lifetime experiment to achieve a precision at the level of 10^{-3}, the available and planned UCN sources at which one might perform such an experiment, and some of the considerations for the guide system which can impact the density to which the experiment is loaded. In the case of UCNτ, we expect that an experiment at this level could be reasonably be performed at the Area B source at LANL.

Acknowledgements

This work was supported by NSF grant number 1005233 and DOE grant number DE-FG02-97ER41042. We also acknowledge the very rapid and thoughtful responses to our requests for summary information on each current and planned UCN source by the scientists and staff working with these sources.

References

1. A. Serebrov *et al.*, Phys. Lett. **605**, 72 (2005).
2. J. M. Doyle and S. K. Lamoreaux, Europhys. Lett. **26**, 253 (1994).
3. A. Serebrov *et al.*, Phys. Rev. C **78**, 035505 (2008).
4. P. L. Walstrom *et al.*, Nucl. Instrum. & Meth. in Phys. Res. A **559**, 82 (2009).
5. A. Steyerl *et al.*, Phys. Rev. C **86**, 065501 (2012).
6. A. Pichlmaier *et al.*, Phys. Lett. B **693**, 221 (2010).

7. A. Saunders *et al.*, Phys. Lett. **B 593**, 55 (2004).

8. A. Saunders *et al.*, Rev. Sci. Instrum. **84**, 013304 (2013).

9. A. Steyerl *et. al*, Phys. Lett. A **116**, 347 (1986) and private communication with A. Steyerl and P. Geltenbort (logbook data showing 39 UCN/cm^3 detected in bottle and production rates).

10. O. Zimmer *et al.*, Phys. Rev. Lett. **107**, 134801 (2011).

11. B. Lauss for the UCN Project Team, AIP Conference Proceedings **1441**, 576 (2012) and private communication with K. Kirch (2012).

12. Y. Masuda *et al.*, Phys. Rev. Lett. **108**, 134801 (2012).

13. T. Lauer, private communication (2013).

14. E. Korobkina *et al.*, Nucl. Instrum. and Meth. in Phys. Res. A **579**, 530 (2007).

15. A. Frei, private communication (2012).

16. J. Martin, private communication (2012).

17. A. Serebrov *et al.*, Physics of the Solid State **52**, 1034 (2010) and an update from A. Serebrov, "PNPI-ILL-PTI collaboration on the nEDM at the ILL reactor in Grenoble and the WWR-M reactor in Gatchina," Oak Ridge nEDM workshop, October 13th, (2012) – presented by P. Geltenbort.

Phase Space Evolution in Neutron Traps for Measurements of the Neutron Beta-Decay Lifetime

C.-Y. LIU*, D. SALVAT and E. ADAMEK

*Center for Exploration of Matter and Energy,
Physics Department, Indiana University, Bloomington, IN 47408, USA*
*CL21@indiana.edu

In trap-based lifetime experiments, the key to extrapolating the neutron β-decay rate is the understanding of non-β-decay losses of the ultracold neutron (UCN) population in the trap. Use of a magnetic trap eliminates the potential for UCN to be lost at surface boundaries. However, these traps also introduce additional systematic errors, such as spin-flip loss when neutrons cross regions of zero field. In addition, the NIST lifetime experiment reported the unexpected presence of quasi-bound, high-energy neutrons that significantly reduced the measured storage lifetime. We discuss the precision required in measuring these sources of non-β-decay losses and strategies to mitigate some of these effects. The discussion will focus on the magneto-gravitational trap used in the UCNτ experiment.

Keywords: Ultra-cold neutron; neutron lifetime; β decay; magnetic trap

1. Introduction

Eighty years after Chadwick discovered the neutron,[1] today physicists still cannot agree on how long the neutron lives. There are two ways to measure the neutron β-decay mean lifetime. One technique looks at neutron decays in-flight, where the lifetime is inferred from the ratio of the rate of the emerging β-decay products (protons or electrons) and the fluence of the incident cold neutron beam. The other technique involves trapping neutrons in spatial confinement, where the lifetime is measured by monitoring the exponential decay of the neutron population in the trap. In the past decade, the availability in the intense sources of ultracold neutrons (UCN) has triggered rapid progress on many experiments that use trapped UCN, including the measurement of the neutron lifetime. UCN are low-energy neutrons that are strongly influenced by material potential and gravitational and magnetic fields. Neutrons are trapped using one or a combination of these interactions. Here, we will focus on trap-based techniques that use UCN and critique their technical feasibilities and systematic errors. Specifically, I will examine the dynamics of trapped UCN in the phase space and discuss how they give new insights into interpretations of experiments.

We start by reviewing how a trap-based experiment typically proceeds. Take the MAMBO experiment[2,3] as an example: UCN are guided from the production source into the material trap, whose surface is carefully prepared and coated with Fomblin oil (with very low probability for neutron loss). When the UCN population saturates in the trap, the entrance and exit valves are closed, leaving the UCN free to decay and to sample the trap volume that is bounded by material walls. After a predetermined storage time, the exit valve opens and surviving UCN are drained into an external UCN detector and counted. We call this method "fill & dump". One can also perform real-time counting of either protons or βs[4] to directly measure the rate of β-decay; however, the high background rate due to γ-induced Compton scattering (for β detection) and the need for an HV-biased detector (for proton detection) make this technique more challenging.

In trap-based experiments, the measured observable is the storage time, $\tau_{storage}$, which is the inverse of the UCN disappearance rate. In addition to β-decay, UCN disappear from the trap due to both nuclear absorption and upscattering loss, when they interact with residual gas molecules in the trap vacuum and material walls on the trap boundary. They also disappear because of heating loss from quasi-elastic scattering off material walls or time-dependent variation of fields in a magnetic trap. In magnetic traps, they can suffer additional loss due to spin-flip when crossing field-zeros and subsequent escape. There is also a slow leaking of the quasi-bound UCN (discussed later in more detail) and possibly many other subtle sources of loss yet to be discovered. To the zero-th order, the rate of UCN disappearance is a linear sum of all loss rates mentioned above:

$$\frac{1}{\tau_{storage}} = \frac{1}{\tau_{\beta}} + \frac{1}{\tau_{abs}} + \frac{1}{\tau_{up}} + \frac{1}{\tau_{sf}} + \frac{1}{\tau_{heat}} + \frac{1}{\tau_{qb}} + \dots. \tag{1}$$

The $\tau_{storage}$ is, thus, systematically shorter than τ_{β}. Many of these processes involve excitations in condensed matter and atomic transitions, which are of comparable time scale as the neutron β-decay ($\tau_{\beta} = 880.0 \pm 0.9\ s$[5]). Dubbers and Schmidt[6] accentuated the challenge of this technique by pointing out that in prior material trap experiments, corrections as large as 150 s were made on the measured storage time in order to infer the neutron β-decay lifetime. In the most precise trap-based experiment,[7,8] a very large material trap ($\approx 300\ l$) was used to reduce the probability of surface interactions, and the correction was successfully reduced to 1 s. In addition, the storage times of the UCN with several velocity groups were independently measured in two different-sized traps. The results were analyzed by parameterizing the non-β-decay loss using one single loss parameter γ, and the experimenters reported the neutron β-decay lifetime by extrapolating the storage time to $\gamma = 0$. This lifetime was 8 s shorter than the PDG average at the time of publication in 2005. This result aroused much contention among experts and has had a far-reaching impact on nuclear, astro-, solar, and neutrino physics, for many of these fields require precise knowledge of the neutron lifetime as an input parameter.

2. Controlling Non-β-decay Losses

Due to the ephemeral nature of the neutron, determining the β-decay lifetime τ_β to better than 0.1% precision requires a heroic effort to carry out a comprehensive measurement program that can quantify all possible losses in a trapped UCN population. Before delving into details of experimental techniques, one should ask a practical question: How well does one have to measure non-β-decay losses in order to achieve a given precision, ϵ, in determining τ_β? To estimate the required precision, we take the derivative of eqn.(1):

$$\frac{1}{\tau_{storage}}\left(\frac{d\tau_{storage}}{\tau_{storage}}\right) = \frac{1}{\tau_\beta}\left(\frac{d\tau_\beta}{\tau_\beta}\right) + \frac{1}{\tau_{abs}}\left(\frac{d\tau_{abs}}{\tau_{abs}}\right) + \frac{1}{\tau_{up}}\left(\frac{d\tau_{up}}{\tau_{up}}\right) + \ldots \quad (2)$$

The relative uncertainty of τ_β emerges from the above equation as:

$$\left(\frac{d\tau_\beta}{\tau_\beta}\right) = \left(\frac{d\tau_{storage}}{\tau_{storage}}\right)\frac{\tau_\beta}{\tau_{storage}} - \left(\frac{d\tau_{abs}}{\tau_{abs}}\right)\frac{\tau_\beta}{\tau_{abs}} - \left(\frac{d\tau_{up}}{\tau_{up}}\right)\frac{\tau_\beta}{\tau_{up}} - \ldots \quad (3)$$

The loss terms are presumably independent of each other, with each representing a new degree of freedom in the problem. Therefore, the relative uncertainty in τ_β is, in fact, a vector sum of the relative uncertainty of each loss term, weighted by the ratio of τ_β to the loss lifetime. As such, we take the modulus

$$\left(\frac{d\tau_\beta}{\tau_\beta}\right) = \sqrt{\left[\left(\frac{d\tau_{storage}}{\tau_{storage}}\right)\frac{\tau_\beta}{\tau_{storage}}\right]^2 + \left[\left(\frac{d\tau_{abs}}{\tau_{abs}}\right)\frac{\tau_\beta}{\tau_{abs}}\right]^2 + \left[\left(\frac{d\tau_{up}}{\tau_{up}}\right)\frac{\tau_\beta}{\tau_{up}}\right]^2 + \ldots}$$
$$(4)$$

as an estimate of the total (relative) uncertainty in τ_β. It is not surprising to find that the contribution of each loss term to the total uncertainty is directly proportional to its loss rate. In other words, the longer the loss time, the less sensitive τ_β is to the fluctuation of that particular source of loss. Many of the loss rates are in fact correlated (most likely through dependence on the UCN energy) and may not be measured individually. Since the initial energy spectra of the UCN is usually not known to tight precision, a multi-variable analysis on multiple measurements with detailed Monte-Carlo simulations may be required to disentangle each loss term. Nevertheless, eqn.(4) provides a good starting point. In order to measure the β-decay lifetime to a relative precision of ϵ, we need to control each loss term to a tenth of ϵ, so that the uncertainty remains statistically limited. Take the absorption loss as an example: the criteria

$$\left(\frac{d\tau_{abs}}{\tau_{abs}}\right)\frac{\tau_\beta}{\tau_{abs}} = \epsilon \times \frac{1}{10} \tag{5}$$

sets the budget for this particular systematic error. New experiments have storage times approaching the β-decay time ($\tau_{storage} \approx \tau_\beta$) and the correction $\Delta t = \tau_\beta - \tau_{storage}$ is quite small. We can express eqn.(1) in terms of Δt:

$$\frac{1}{\tau_{storage}} - \frac{1}{\tau_\beta} = \frac{\Delta t}{\tau_\beta \cdot \tau_{storage}} = \frac{1}{\tau_{abs}} + \ldots . \tag{6}$$

If other losses are small in comparison, the maximum contribution from the absorption loss is, thus,

$$\left(\frac{\tau_\beta}{\tau_{abs}}\right)_{max} = \frac{\Delta t}{\tau_{storage}} \approx \frac{\Delta t}{\tau_\beta}, \tag{7}$$

which can be put into eqn.(5) to yield an estimate of the required precision for the absorption loss:

$$\left(\frac{d\tau_{abs}}{\tau_{abs}}\right)_{min} = \frac{\epsilon}{10} \times \frac{\tau_\beta}{\Delta t}. \tag{8}$$

In a low-loss trap, the correction time $\Delta t \approx 1$ s has been demonstrated. According to eqn.(8), in order to attain $\epsilon = 0.1\%$ precision in τ_β, one has to measure the absorption loss to a 10% accuracy. In this situation, the characteristic time of the absorption loss is $(\tau_\beta \tau_{storage})/\Delta t \approx \tau_\beta^2/\Delta t \approx 10^6$ s, which is about 10 days. The fact that the absorption time is long does not require neutrons to be stored for this impossibly long duration. Instead, quantifying this long loss time would involve measuring changes in the storage time while systematically varying the absorption rate, for instance, by raising the partial pressure of residual gas. It can be done provided large statistics on UCN counting with a good signal-to-noise ratio could be achieved. In the other scenario, when the correction Δt is large, the loss rate is comparable to the β-decay rate. If $\Delta t \approx 100$ s, then eqn.(8)

dictates that we measure the loss to the same desired precision of β-decay to 0.1%. Both scenarios require dedicated time to measure the loss, but the latter faces a bigger challenge to achieving a precise understanding of non-β-decay losses.

3. Qausi-Bound Neutrons

To date, the dominating systematic effect faced by trap-based experiments has been neutron loss from material interactions. Surface loss has been shown to be difficult to model and hard to quantify experimentally (partly due to the unknown UCN energy spectrum).[9-11] To avoid this effect, many recent experiments have begun to explore magnetic traps that confine neutrons of one spin state.[4,12,13] The so-called "low-field seekers" deflect from strong magnetic fields and can be trapped using a simple quadrupole magnet. A moderate field strength (of 6 T) is required to fully trap UCN with energy up to 350 neV. For example, the NIST UCN lifetime experiment demonstrated the magnetic trapping of the UCN produced in superfluid helium.[4,14]

The NIST UCN experiment made many technical advances, including the superthermal production of UCN using Bragg-scattered mono-energetic 9 Å neutrons and the detection of β-decay events through scintillations in superfluid helium. It also revealed the presence of "marginally" trapped neutrons. These neutrons have kinetic energy larger than the trapping potential, but the excess kinetic energy is stored in motion tangential to the radial trapping force. Their radial kinetic energy is not large enough to overcome the trapping potential, and the neutron is bound in the trap indefinitely. As such, these neutrons carry large angular momentum. If they remain in stable orbits, these neutrons should have no effect on the lifetime measurements, as they undergo the same β-decay as their low-energy counterparts. Unfortunately, the presence of small perturbations spoils the rotational symmetry, and relaxes the law of angular momentum conservation. The perturbations lead to a small degree of orbit-mixing. The consequence is a slow conversion of tangential kinetic energy to radial motion, leading to a slow escape of the UCN from the trap. We prefer to call them "quasi-bound" neutrons, in accordance with the description provided here.

The rate of conversion is determined by the strength of perturbation, which can exist in the form of fringe fields, high-order multi-poles, imperfect symmetry of the constructed trap, and the gravitational field (if not aligned with the symmetry axis of the magnetic field). The escape of quasi-bound neutrons can lead to a false interpretation of the β-decay lifetime. In order to mitigate this problem, the NIST experimenters added a step to purge these quasi-bound neutrons by ramping down the magnetic field and lowering the depth of the trapping potential to 30%. The field was then ramped back up to full strength of 1.1 T to begin storage. A 50% reduction in the neutron number was the consequence of this attempt to control this systematic effect.

Compared to the field in magnetic traps, the material potential is short-range, rising sharply only when neutrons come in close vicinity to the trap walls. If the process of neutrons scattering off material surfaces is highly specular (i.e., the scattering does not randomize the direction of the reflected UCN), then high-energy neutrons can preserve their large angular momentum for a very long time in the trap. This phenomena in material traps present the same plausible problem of quasi-bound neutrons as observed in the magnetic trap at NIST. The escape of quasi-bound neutrons from the trap opens up another channel of neutron loss, adding a $1/\tau_{qb}$ term in eqn.(1).

The escape time depends upon many details, including the energy spectrum of the trapped neutrons, the angular distribution of the neutrons when entering the trap, details of the trap geometry and trapping potential, the surface roughness in the case of material traps, and the field ripples in the case of magnetic traps. Unless a large-area detector is deployed to monitor the escaped neutrons, the corrections due to the quasi-bound neutrons have to be studied via modeling using Monte-Carlo simulations. In material traps, however, the problem might be somewhat alleviated because neutron scattering off surfaces tends to have a significant diffusive component. In order to achieve the desired precision of 0.1% and beyond, the dynamics of the trapped neutrons deserve further investigation.

4. Neutron Dynamics in the Phase Space

A trapped neutron is localized in space and its energy is bound by the depth of the trapping potential. The confinement of neutrons is, thus, best illustrated in the phase space, where each neutron traces out a trajectory in this 6-dimensional hyperspace. In some cases the trajectories might be closed and periodic, but in general they are not. Due to the deterministic nature of classical mechanics, the trajectories in the phase space do not intersect. If the potential is time-independent (autonomous Hamiltonian), the trajectories in the phase space are invariant. Furthermore, with conservative forces, Liouville's theorem states that the phase space volume of a statistical ensemble of particles is invariant. Thus, if the particle number is conserved (i.e., no β-decay) then the particle flow in the phase space can be treated as an incompressible fluid.[15]

We will attempt to understand the lifetime measurement procedures by analyzing how neutrons flow in the phase space. The sequence of a measurement cycle in a trap-based experiment typically consists of the following four steps: Fill, Spectrum cleaning, Storage, and Detection. In the "Fill" step, the phase space is populated by neutrons from a source. In the location of the trap, neutron density is concentrated as a result of the trapping potential. Due to the broad energy distribution of the incoming neutrons and the finite size of the trap, the initial ensemble of neutrons occupies a finite, closed volume in the phase space. The initial filling of neutrons in the phase space is unlikely to be uniform; nevertheless, we expect it to reach a static equilibrium (as the potential during the Fill is time-independent). The time it takes to establish equilibrium sets the time needed for the Filling. In Cartesian coordinates, this phase space volume remains localized in the phase space. In coordinates using angle-rotation, however, the phase space volume would start to advance in ϕ as particles orbit.

Trapped neutrons trace out continuous trajectories that are bound within a closed phase space volume. In contrast, un-trapped neutrons occupy an adjacent volume with higher momentum, and over time they would flow out of the spatial boundaries of the trap. In order to eliminate contamination from quasi-bound neutrons, the goal is a total purge of the neutrons over the energy threshold. The subsequent step of "Spectrum Cleaning" can be conducted by two alternate methods. The first uses a neutron absorber that is brought into the trap momentarily. The position and energy response function of the absorber trace out a loss boundary (as a hyper-surface) on the phase space. High-energy neutrons are absorbed when their trajectories cross this loss boundary. The second method involves ramping the trapping potential. As the magnetic field is ramped down, the phase space volume of the trapped neutrons morphs (but the total volume does not expand) until the outlier neutrons cross the loss boundary defined by, for example, the neutron-absorbing surface inside the cryostat. Each cleaning mechanism draws a different loss boundary. Whether the trap is cleaned sufficiently depends how the phase space volume intersects the loss boundaries and how long the phase space volume is allowed to evolve.

The lifetime measurement takes place during the next step of "Storage". In the phase space, each neutron traces out a non-intersecting trajectory that is invariant in time; the neutrons stay close together, and the phase space volume they occupy remains connected. Experimentally, the trap is best defined by spatial boundaries. The energy boundary can often only be applied to certain components of momentum because of the directionality of trapping fields. It is therefore harder to draw an energy boundary as the momentum can be converted from one form to another (as discussed before). Within the spatial boundary of the trap, the neutron population will decrease due to β-decay. In the meantime, absorption and upscattering (with residual gas molecules and material walls) also depopulate neutrons following a characteristic $1/v$ dependence. The two loss rates also depend on the collisional frequency that is different for low and high-energy neutrons. Furthermore, neutrons undergo spin-flips and escape the magnetic confinement when their trajectories cross regions of the zero-field. These mechanisms of loss present as sinks of neutrons in the phase space. Some are localized in the phase space, while others have continuous response functions over a wide range in position and momentum. If there exist sources of neutron heating that pump energy into the neutron ensemble, over time the closed phase space volume will grow, making some neutrons cross the loss boundary. The source of heating can often can be described using time-varying potentials. All of these sources of non-β-decay loss will need to be controlled and measured.

In the final step of "Detection", a sink is put into the phase space so that the neutrons can be directed into a detector and counted. Details of detection vary for different experiments. In the "fill & dump" method, an opening is introduced in the trapping potential in order to allow neutrons to flow into a detector placed outside the trap. Depending on the size of the opening, it might take some time for the neutrons to be completely drained out of the trap and counted. There is also the question of whether some of the neutrons will stay in the trap indefinitely, as they are stuck in orbits that do not intersect with the detector sink. If so, how big is the population of these neutrons?

All of these phenomena generalize into a question we can ask using the phase space terminology: to successfully extract the β-decay lifetime, we need to determine whether the phase space volume defined by the spectrum-cleaned ensemble of neutrons stays invariant throughout the closed-door storage. Does the total volume of phase space volume change? Does the contour of the volume change? The closed phase space volume occupied by the trapped neutrons flows in 6-dimensional hyperspace, constantly changing its shape as guided by the Hamiltonian equations. Could the closed phase space volume evolve and intersect our loss boundary in some unexpected way? We shall come back to this question in the following section.

5. Dynamics of the UCN in the UCNτ Apparatus

The character of a trap is defined by its trapping potential. A Hamiltonian with N degrees of freedom is integrable, if and only if, there exist N independent isolating integrals. These integrals are action variables, which form the global invariants of the motion. This Hamiltonian produces regular, closed orbits. On the other hand, a near-integrable potential lacks invariant action variables due to somewhat-spoiled global symmetries. The near-integrable potential produces mixed dynamics, with regular, quasi-periodic orbits separating regions of stochasticity. The regions of regular and stochasticity are separated by Kolmogorov-Arnold-Moser (KAM) curves that are well studied in one- and two-dimensional systems.[15] These KAM curves characterize the persistent quasi-periodic motions under small perturbations. For dimensions larger than two, the region of stochasticity often forms a "web" of regions that are connected. Particles flow between regions through Arnold diffusion.[15] For most UCN traps (with spherical, cylindrical or cubic geometries), the potential is probably near-integrable, due to tolerances in machining and alignment.

For the rest of the discussion, we will use the trap geometry of the UCNτ experiment in order to illustrate the neutron dynamic in phase space. The UCNτ apparatus is a hybrid magneto-gravitational trap.[16,17] It uses a large Halbach magnet array with an optimized near-field gradient to provide a magnetic levitating force. The magnet array is similar to a large household bathtub. The top of the trap is free of materials and magnets, as the UCN are bound vertically by the gravitational field on the surface of the Earth. For a neutron, the energy required to climb a height of 0.5 m matches the coupling energy to a magnetic field of 0.8 T. The trapping force comes from the magnetic moment of the neutron reacting to the local magnetic gradient. A Halbach configuration requires four magnets to complete a full rotation of the magnetization vector. The resulting magnetic field decays exponentially from the magnet pole surface with a characteristic length of four times the size of the individual permanent magnet. The large volume of this trap (≈ 670 l) increases the statistics of trapped neutrons and β-decay events; the open-top geometry allows ample room for neutron detection inside the trap.

Unlike the previous UCN traps, this trap is designed in such a way as to break rotational symmetry by trapping UCN inside a vertically-placed toroid, characterized by its different major R and minor r radii. The global symmetry is further broken by piecing together two toroids of the same R, but different r. The left half is a ring toroid (with $r_L < R$) and the right

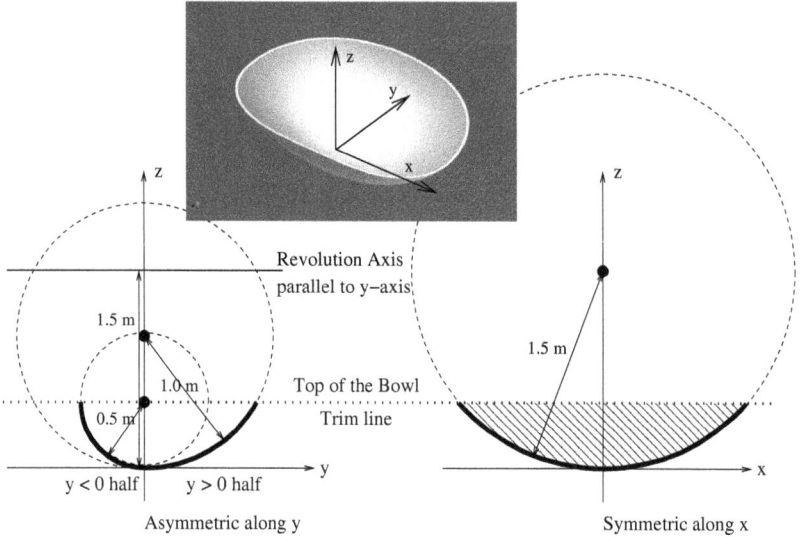

Fig. 1. A schematic of the asymmetric toroidal design of the magnet array. In the global coordinate system, the x axis is along the symmetric axis of the toroid (with the major radius $R = 1.0$ m), while the y axis is along the asymmetric axis (with different minor radii, $r_R = 1.0$ m and $r_L = 0.5$ m) and the z axis points up vertically. The thick solid curves represent the inner-surface profile of the Halbach array (Color Plate 2).

half is a spindle toroid (with $r_R > R$). Note that the geometry can be morphed into a degenerate sphere when $r_L = r_R = R$. We will take advantage of this feature in the following simulation studies. Since the vertical motion of the UCN is bound by gravity, they are concentrated on the bottom of this vertically-oriented toroidal geometry. As such, only the bottom of the toroid needs to be constructed to form a bowl (or bathtub) shape as shown in Fig.(1). Deliberately breaking the rotational symmetry avoids the stable quasi-bound orbits with large angular momentum. Introducing large perturbations in the trap symmetry should lead to rapid orbit-mixing, allowing quasi-bound neutrons to sample a large portion of the phase space. This increases the probability of UCN crossing loss boundaries and thus makes them escape the trap quickly. This asymmetric arrangement of the magnet array, together with field ripples resulting from high-order multipole expansions of discrete elements of permanent magnets, should render the trap "self-cleaning". The motivation to invoke chaotic dynamics of trapped UCN, proposed initially by Bowman,[18] was to accelerate the cleaning process to purge quasi-bound neutrons out of storage. This added feature should lessen the reliance on field ramping and neutron absorbers

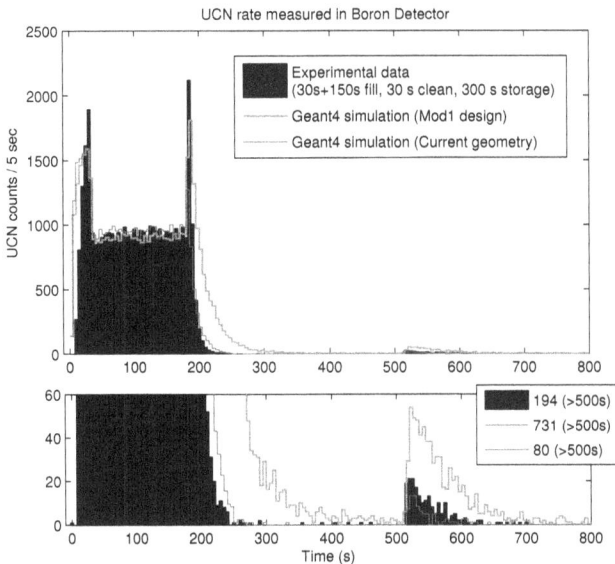

Fig. 2. A comparison between the GEANT4 simulated timing spectra and the measurement in the "fill & dump" mode (blue histogram: 180 s filling, 30 s cleaning and 300 s storage.). The magenta curve is the simulated result using current geometry, while the green curve is the predicted enhancement with a MOD1 design, which is used to increase the UCN loading efficiency into the trap (Color Plate 2).

as the only means of purging quasi-bound neutrons.

The Halbach magnet array was finished in 2012; shortly after, the experiment took first data with UCN at the Los Alamos Neutron Science Center (LANSCE) in early 2013. Using this apparatus, with one week of beam time, we have observed UCN trapping and quantified the UCN storage time. The data analysis has been completed and planned improvements on the hardware are underway. In order to help us understand the details of the experiment, we have built full-scale simulations to track the UCN and study the response functions of this apparatus. The observables that we can compare directly to the measurements include the time scale for Filling, Cleaning, relative count rate between the UCN monitors, and time-of-flight spectra of the UCN in the process of neutron draining into an external UCN monitor. For instance, Fig.(2) shows the simulated result (built on Geant4UCN[19]) on our boron-coated detector, which serves as an external UCN detector.

We have also developed several numerical simulations to study long-term tracking of the UCN inside the trap. Since the motion is governed by Hamiltonian mechanics, the simulation employs simplectic integrators in order to attain long-term stability and enforce energy conservation. Fig.(3) shows a simulated result on the position distribution of the neutron absorption on a large sheet of the UCN spectrum cleaner covering the top of the trap. Depending on the height of the absorber, the quasi-bound neutrons can be cleaned to 99% over 50 s. As shown here, for quasi-bound neutrons with energy up to 10% higher than the trap potential ($V_0 = 50$ neV), the distribution of the neutron absorption events on the UCN cleaner is far from uniform. We can attempt to understand this distribution by plotting a typical trajectory of the UCN inside the trap, as shown in Fig.(3). Note that when it is near the bottom of the trap, the horizontal motion of the neutron is converted to vertical motion as it climbs up the side of the trap bowl. This effect is similar to the phenomena that when light rays emitted from a source placed near a focal point become parallel upon reflection from a concave mirror, or similar to a Winston cone for neutrons. The important lesson learned here is that we cannot assume each neutron in a magnetic trap will uniformly and quickly access the whole kinematically-allowed volume in the phase space or in the trap. The neutron trajectories are far from being ergodic. Even though the quasi-bound neutrons have enough energy to reach the spectrum cleaner, if a practical cleaner (of a much reduced

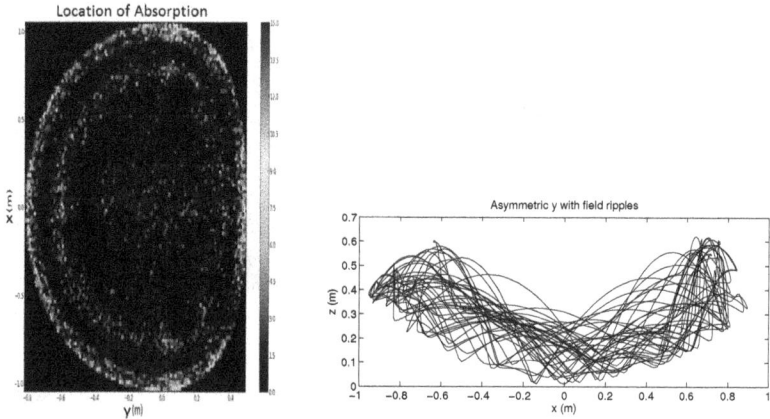

Fig. 3.   *Left*: A 2-D histogram of the absorption events of the high-energy UCN (with $E = V_0 + 6$ neV) on a UCN absorber, fully covering the top of the trap. *Right*: A example 3-D trajectory of a high-energy UCN inside the trap, numerically integrated using a simplectic integrator of the full Halbach magnetic potential (on the asymmetric toroid with field ripples) (Color Plate 3).

size) is placed in the center of the trap, we could miss a large portion of the quasi-bound UCN, resulting in a poor efficiency of Spectrum Cleaning.

5.1. *Poincaré's Section of Surface*

In order to categorize the types of motion inside the trap, we adapted the method of Section of Surface developed by Poincaré. To gain some quick insight into the problem without resorting to the full power of computation using realistic fields, we applied some simplifications to the trapping fields. After reviewing the UCN dynamics in the idealized geometry, we will remove these simplifications step-by-step in order to observe how the dynamics of the UCN evolve with each level of added complication (or perturbations on the global symmetry). To start, we will approximate the Halbach magnetic gradient as a perfect mirror, ignoring the dispersion resulting from the finite-range of magnetic potential. We will also ignore the field ripples. In reality, the amplitude of the field ripple experienced by neutrons is energy dependent: the higher the energy carried by a neutron, the deeper it can penetrate into high field regions with stronger ripples. In this simplification, when a UCN is reflected from the magnet pole surface, its velocity component along the surface normal is reversed, and the tangential component is preserved. After it has been reflected, the UCN motion is affected by gravity alone. Given the position and reflected velocity, we can quickly find the next point of UCN collision on the trap surface by solving for the intersection point of the parabolic trajectory on the toroidal surface. Since it is difficult to visualize the dynamic of UCN in the 6-dimensional phase space, we follow Poincaré's prescriptions to map the UCN motions on the Section of Surface.[20]

In this simplified model, the mapping between one point of collision and the next is unique, as the motion between each surface landing follows the simple kinematics of constant acceleration. Without numerically integrating the parabolic trajectories of UCN inside the trap volume, we only need to record x-p information for the collisional points. The time between the subsequent crossings of the Section of Surface varies, but it can be easily worked out if needed. We start with a degenerate sphere by setting $r = R = 0.5$ m. Here, the neutron trajectories are confined on a 2-D vertical plane since the system has rotational symmetry around the z-axis, where the gravity is aligned. In Fig.(4), we plot a Poincaré map of radial velocity, v_r, and transverse velocity, v_θ, upon every wall collision. The radial and transverse velocities are obtained by taking the projection relative to the local surface normal vector:

$$v_r = \vec{v} \cdot \vec{n}; \qquad v_\theta = \sqrt{v^2 - v_r^2}. \qquad (9)$$

In order to simulate the Filling step of the lifetime experiment, mono-energetic neutrons are generated at the very bottom of the trap. Following the history of each neutron, we can trace out a solid curve on the Poincaré's Section of Surface. Each neutron with a different velocity vector traces out a different continuous curve. The fact that all crossing points lie on a curve gives explicit demonstration of the algebraic connection between the two free degrees of freedom, as manifested by the working of the trapping field. The incident polar angle θ (i.e., the angle sustained between the velocity vector and z-axis) is scanned from 0 to 90°, corresponding to shooting vertically up and rolling along the surface, respectively. As shown in Fig.(4), for $v = 1.5$ m/s, the curves for each incident angle are well-defined and nicely separated. All of the curves satisfy of energy conservation:

$$\text{KE} = \frac{1}{2}mv_r^2 + \frac{1}{2}mv_\theta^2 = C - mgz \qquad \longrightarrow \qquad \text{KE}^{max} = C, \qquad (10)$$

where C is a constant of total energy and z is the vertical height on the point of reflection measured from the bottom of the trap. The history of this 1.5 m/s neutron encloses half of an ellipse, as v_r is derived from the reflected velocity and stays positively defined. For a neutron entering at a large polar angle (or a small glazing angle to the surface), its vertical velocity is small and the accessible range for the tangential velocity is large. These neutrons are rolling up and down the surface of the bowl, and the spatial volume they occupy is confined close to the surface. For a neutron with a small polar angle, the vertical reach of the neutron is large as it's velocity is initially vertical, but the points of landing on the trap surface are narrowly confined to the bottom of the bowl. Due to the small slope on the surface near the bottom of the trap, the initially-large vertical velocity is never converted into the horizontal velocity needed to move along the curved surface. These all constitute regular orbits as solutions of deterministic classical mechanics. Again, the actual volume occupied by each neutron is much smaller than the total volume of the trap. This observation of reduced accessible volume will have important implications on the estimate of the achievable UCN density and distribution in the trap.

As the velocity of the UCN is increased to 2.0 m/s, the envelope allowed by energy conservation grows larger as expected. Regular orbits still exist for neutrons entering vertically at a small polar angles, but the kinematic curves on the Poincaré map start to develop characteristic features resem-

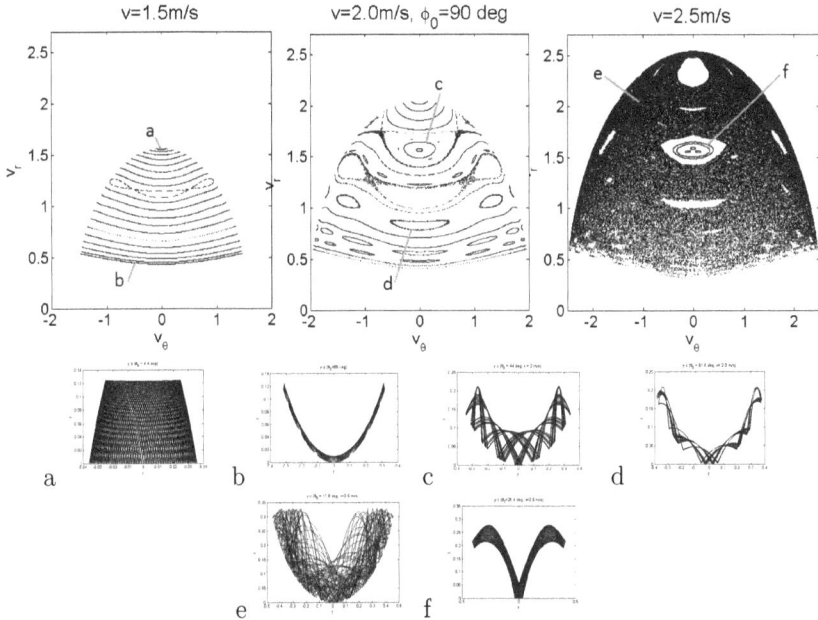

Fig. 4. *Top row*: Poincaré maps of the UCN dynamics in a degenerate sphere with $r = 1.0$ m. The curves on each plot have the same velocity, but each curve represents a polar angle incremented by 4.4° from $\theta = 4.4°$ (top curve) to 88° (bottom curve). *Bottom row*: 2-D UCN trajectories corresponding to the labeled points on the Poincaré maps.

bling the onset of chaotic dynamics in many non-linear systems. Islands of stability develop where the regular orbits reside, separated by regions of stochasticity. When the motion becomes stochastic, the points on the Section of Surface are no longer connected to make a smooth curve. Locations of the stability islands also follow regular patterns that are repeated on finer and finer scales. Sample 2-D trajectories that correspond to some points on the Poincaré map are plotted in Fig.(4). Stochastic regions expand as the UCN velocity is increased beyond 2.0 m/s. At $v = 2.5$ m/s, the map of the UCN dynamic on the Poincaré's Section of Surface looks like Fig.(4). A few prominent islands and many smaller ones exist toward the bottom in a vast sea of stochasticity. Even in the region of stochasticity, the neutron dynamics are not totally ergodic, i.e., the phase space accessible to the UCN with a particular kinematic condition is limited to a certain sub-region (as shown by an example of a 2-D trajectory in Fig.(4)). The overlap of the sub-space makes up the distribution shown in Fig.(4).

In this 2-D system, the UCN dynamics can also be described by other kinematical variables in addition to v_r and v_θ. Plotting z-v_θ from the UCN trajectories studied in Fig.(4) on the Section of Surface gives us Fig.(5). Now, the envelope determined by energy conservation is described by

$$\frac{1}{2}mv_\theta^2 + mgz = C - \frac{1}{2}mv_r^2. \tag{11}$$

Not surprisingly, the information presented in the new plot is topologically identical to that shown previously because the total degree of freedom of the system is only 3. The points of large v_r in Fig.(4) are mapped to points of the small z in Fig.(5), and the points of the small v_r in Fig.(4) now map to the region close to the outer contour of the map in Fig.(5). Following the discussions in,[21] which studied gravitational billiards on an integrable parabolic surface, we conclude that the onset of chaos is correlated with whether the UCN has enough energy to reach the height of the focal point. On a circular concave surface with a radius of curvature r, the focal point is at $\approx r/2 = 0.25$ m from the surface. Note that the z in Fig.(5) is the height of the UCN landing on the trap surface, as such, the maximum height that can be reached by the UCN is higher than the z in the figure.

In order to extend the motion from 2- to 3-D, we set $r = 0.5$ m and $R = 1.0$ m on our toroidal trap and allowed the initial UCN velocity to point out of the symmetry planes. The azimuthal angle of the initial UCN velocity is varied from 0 to 90°, corresponding to pointing along the plane of the major and minor radii, respectively. Along $\phi = 0$, the UCN motion is again confined to a 2-D plane, which contains the larger radius of curvature R, as it is a symmetry plane. Here, even with $v = 2.5$ m/s, the maximum speed

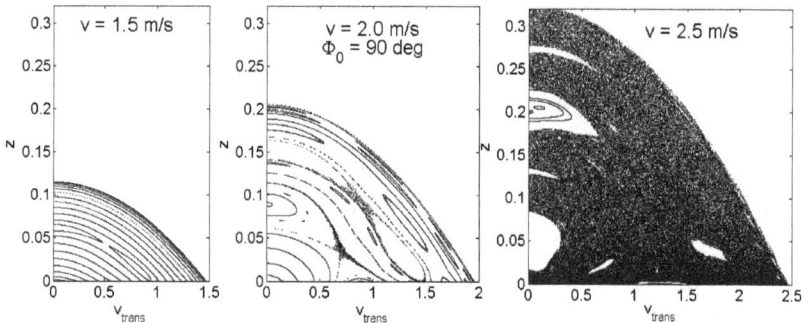

Fig. 5. Poincaré maps plotting v_θ and z at the points of landing on the surface of the trap bowl. The geometry of the trap is the same degenerate sphere used in Fig.4. The transverse velocity is $v_{trans} = v_\theta$ (Color Plate 3).

investigated in the previous case, all the orbits are regular, as evidenced by the discrete curves for all polar angles, θ. We expect the motion to be chaotic when the UCN reaches a height of $R/2 = 0.75$ m, which exceeds the height of the trap. As shown in Fig.(6), when a small $\phi = 20°$, the curves of the regular orbits in the previous case expand into stripes on the Section of Surface, and regions between the curves are filled by stochastic events. With an increasing ϕ, the stripes grow bigger, overlapping into a continuum of stochasticity with well-isolated islands of regular, quasi-periodic orbits. As observed here, the neutrons coming in along the minor axis experience more chaotic dynamics than the neutrons coming in along the major axis.

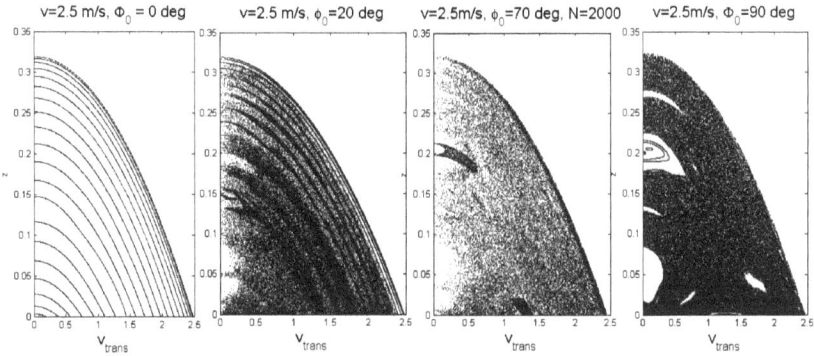

Fig. 6. Poincaré maps of the UCN dynamics in a vertical toroidal trap with a major radius $R = 1.0$ m and a minor radius $r = 0.5$ m. The neutrons are of the same velocity $v = 2.5$ m/s starting at the bottom of the bowl. The incoming polar angle θ is varied the same way as in Fig.4, but the azimuthal angle ϕ is varied from 0 (left graph) to 90° (right graph). The transverse velocity is $v_{trans} = \sqrt{v_\theta^2 + v_\phi^2}$ (Color Plate 4).

Finally, we take the step of making the toroidal bowl asymmetric by allowing the radii (along the minor axis) to differ along the left and right halves of the toroid (i.e., $r_L = 0.5$ m and $r_R = 1.0$ m) while keeping the major radius $R = 1.0$ m. In order to illustrate the effect of this perturbation, we plot the Poincaré map for the UCN with velocity near the onset of chaos on the symmetric toroid, and we compare the maps for the symmetric and asymmetric toroids side-by-side in Fig.(7). The $v_\theta - v_r$ on the Section of Surface is plotted for the $v = 2.0$ m/s projecting along $\phi = 90°$. In the symmetric toroid, thin stochastic regions are bounded by KAM curves. In the asymmetric toroid, the breaking of left-right symmetry introduces a strong enough perturbation to mostly destroy the KAM curves, but there are still regions of the islands for regular orbits.

Another way to illustrate the chaotic dynamic is to plot the time interval between collisions. As shown in Fig.(8), the y-position of the UCN with varying azimuthal angles (at a relatively horizontal incidence, $\theta = 74.8°$) is recorded for each collision up to 2000 collisions. The time between collisions is histogramed to reveal any features in the frequency spectrum. The UCN has a velocity of 2.5 m/s and ϕ_0 of less than $60°$, the motion is semi-periodic, even though the trajectory is not closed, with an average time of 0.1 s between collisions. As an incoming UCN points more toward the minor axis y of the asymmetric toroid, the collisional time distribution broadens, indicating an increasing degree of orbit-mixing. Some trajectories last as long as 0.6 s in free fall before hitting the surface. The directional dependence of the UCN dynamic is consistent with our observations on the Poincaré maps.

Plotting on the Poincaré Section of Surfaces helps to visually categorize UCN dynamics. It serves as an intuitive tool by which to identify chaotic dynamics before working out the Lyopunov exponents. Future studies should continue by adding field ripples and defects into the trap, followed by a full scale simulation with realistic magnetic field potential. To this end, we will summarize what we have learned so far. Chaotic or not, UCN trajectories in the phase space remain invariant as long as the Hamiltonian is time-independent (autonomous). The initial ensemble of the UCN is confined in a well-defined volume in the phase space (i.e., they all emerge from the bot-

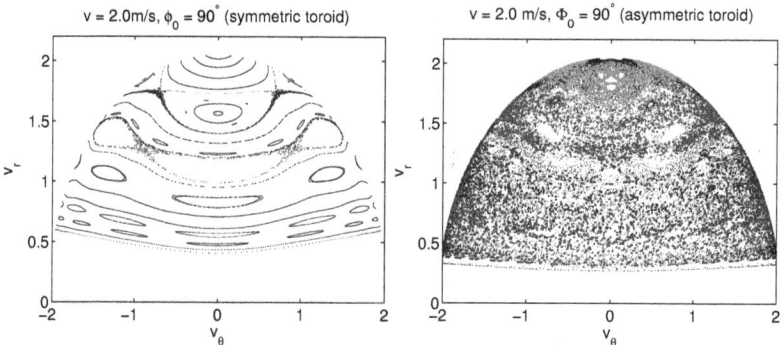

Fig. 7.   Poincaré maps of the UCN dynamics in a symmetric toroidal trap (*left figure*) and an asymmetric toroidal trap (*right figure*), with $R = 1.0$ m, $r_R = 1.0$ m and $r_L = 0.5$ m. These motions are confined to $\phi = 90°$ on a symmetry plane. On the right figure, groups of 4 different θ_0 were highlighted in colors to illustrate the corresponding region of stochasticity (Color Plate 4).

tom of the trap with a proper energy distribution). For low-energy neutrons following regular orbits, the phase space volume occupied by the initial ensemble of UCN will stay connected, as the adjacent points (with a small difference in the initial condition) stay close together and the system occupies a closed volume. The shape of the phase space volume could change, but the total volume is conserved. Since the asymmetric toroidal geometry introduces stochasticity, we deduce that neutrons with energy higher than mgh, where $h = r_L/2 = 0.25$ m, the dynamics in the trap is chaotic. The chaotic dynamic of the high-energy UCN should help with spectrum cleaning in the trap. The phase space volume occupied by these neutrons will evolve and grow filaments as the dynamic undergoes the process of fractalization and expands out to kinematically accessible regions in the phase space. The phase space volume is still conserved, but the coarse-grained volume (by connecting the out-liner events) appear to grow in time. This increase in the coarse-grained volume is in line with the understanding of entropy increase in Hamiltonian systems. The result is that the ensemble of UCN starts to disperse into the phase space. Even though the UCN trajectories are bound in space by the boundary of the trap and bound in momentum by the trap potential, it might take a long time for UCN to explore all kinematically-allowed regions in the phase space.

6. Summary

Although quasi-bound UCN can be cleaned from traps using a UCN absorber, the detailed UCN dynamics in the phase space and their effect on the results have not been seriously considered in most previous experiments.

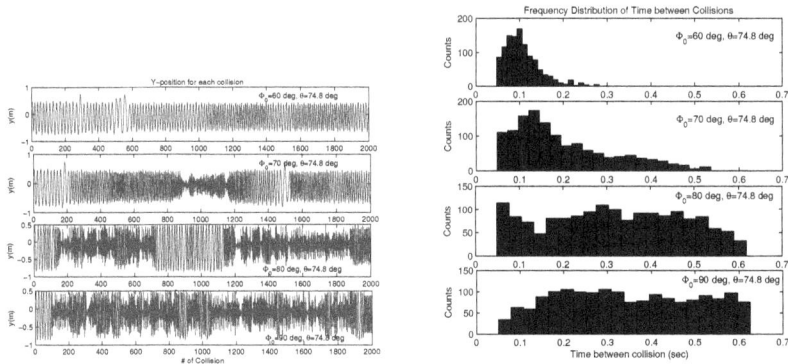

Fig. 8.   History of the collisional position (*left*) and frequency distributions of time between collisions (*right*) (Color Plate 4).

Here, we presented the results of a preliminary study of UCN dynamics in the UCNτ trap. We studied neutron dynamics following standard treatments in the study of gravitational billiards and found that the UCNτ trap presented mixed dynamics, even in the limit of a degenerate sphere with no broken symmetry (except that introduced by gravity). We have shown that for the UCN with energy that is 50% of the trap potential, the Poincaré maps show stochastic distributions in the history of the neutron motion. The degree of the stochasticity is highly dependent upon the incoming angle of the neutrons. We concluded that the quasi-bound neutrons are chaotic in the asymmetric toroidal trap and that the trap appears to be self-cleaning for these high-energy neutrons.

For current and future trap-based lifetime experiments that aim to increase the precision beyond 1 s, subtle details in the phase space dynamic of the trapped UCN could be introduced as systematic errors that need to be monitored and corrected. Knowing how the dynamics evolve in the phase space would help us to construct a more complete physics model to put in the simulation. The interpretation of the β-decay lifetime from the observable storage time relies heavily on simulations. The problem is most severe in the "fill & dump" method, in which almost no information is collected while neutrons are trapped behind a closed door. Simulations, after all, can only be as good as the physics we put in. Thus, we plan to develop physics insights through simulations and, in turn, benchmark the simulation against well-planed diagnostic measurements. This iterative process of cross-checking results of measurements and simulations seems to be the only sensible way to build confidence in our understanding of the trap-based lifetime experiments.

References

1. J. Chadwick, *Proc. R. Soc. London, Ser. A* **136**, 692 (1932).
2. W. Mampe, P. Ageron and C. Bates, *Phys. Rev. Lett.* **63**, p. 593 (1989).
3. A. Pichlmaier, V. Varlamov, K. Schreckenbach and P. Geltenbort, *Phys. Lett. B* **693**, 221(Octtober 2010).
4. P. Huffman, C. Brome, J. Butterworth, K. Coakley, M. Dewey, S. Dzhosyuk, R. Golub, G. Greene, K. Habicht, S. Lamoreaux, C. Mattoni, D. McKinsey, F. Wietfeldt and J. Doyle, *Nature* **403**, 62(January 2000).
5. J. et al Beringer, *Phys. Rev. D* **86**, p. 010001 (2012).
6. D. Dubbers and M. Schmidt, *Rev. Mod. Phys.* **83**, 1111 (2011).
7. A. Serebrov, V. Varlamov, A. Kharitonov, A. Fomin, Y. Pokotilovski, P. Geltenbort, J. Butterworth, I. Krasnoschekova, M. Lasakov, R. Tal'daev, A. Vassiljev and O. Zherebtsov, *Phys. Lett. B* **605**, 72(January 2005).
8. A. Serebrov, V. Varlamov, A. Kharitonov, A. Fomin, Y. Pokotilovski,

P. Geltenbort, I. Krasnoschekova, M. Lasakov, R. Taldaev, A. Vassiljev and O. Zherebtsov, *Phys. Rev. C* **78**, p. 035505(September 2008).

9. S. Lamoreaux and R. Golub, *Phys. Rev. C* **66**, p. 044309(Octtober 2002).

10. A. Serebrov, N. Romanenko, O. Zherebtsov, M. Lasakov, A. Vasiliev, A. Fomin, P. Geltenbort, I. Krasnoshekova, A. Kharitonov and V. Varlamov, *Phys. Lett. A* **335**, 327(February 2005).

11. A. Steyerl, J. M. Pendlebury, C. Kaufman, S. S. Malik and a. M. Desai, *Phys. Rev. C* **85**, p. 065503(June 2012).

12. V. Ezhov, a.Z. Andreev, G. Ban, B. Bazarov, P. Geltenbort, F. Hartman, a.G. Glushkov, M. Groshev, V. Knyazkov, N. Kovrizhnykh, O. Naviliat-Cuncic, G. Krygin, a. Mueller, S. Paul, R. Picker, V. Ryabov, a. Serebrov and O. Zimmer, *Nucl. Instrum. Methods Phys. Res., Sect. A* **611**, 167(December 2009).

13. S. Materne, R. Picker, I. Altarev, H. Angerer, B. Franke, E. Gutsmiedl, F. Hartmann, A. Müller, S. Paul and R. Stoepler, *Nucl. Instrum. Methods Phys. Res., Sect. A* **611**, 176(December 2009).

14. C. Brome, J. Butterworth, S. Dzhosyuk, C. Mattoni, D. McKinsey, J. Doyle, P. Huffman, M. Dewey, F. Wietfeldt, R. Golub, K. Habicht, G. Greene, S. Lamoreaux and K. Coakley, *Phys. Rev. C* **63**, p. 055502(April 2001).

15. A. J. Lichtenberg and M. A. Lieberman, *Regular and Chaotic Dynamics* (Springer-Verlag, 1983).

16. G. Berman, V. Gorshkov and V. Tsifrinovich, *Nucl. Instrum. Methods Phys. Res., Sect. A* **592**, 385(July 2008).

17. P. Walstrom, J. Bowman, S. Penttila, C. Morris and a. Saunders, *Nucl. Instrum. Methods Phys. Res., Sect. A* **599**, 82(February 2009).

18. J. Bowman and S. Penttila, *J. Res. Natl. Inst. Stand. Technol* **110**, 361 (2005).

19. F. Atchison, T. Bryś, M. Daum, P. Fierlinger, A. Fomin, R. Henneck, K. Kirch, M. Kuźniak and A. Pichlmaier, *Nucl. Instrum. Methods Phys. Res., Sect. A* **552**, 513(November 2005).

20. H. Korsch, H. Jodl and T. Hartmann, *Chaos: a program collection for the PC* (Springer, 2007).

21. H. Korsch and J. Lang, *Journal of Physics A* **24**, 45 (1991).

Chaos in a Gravo-Magneto Neutron Trap

J. DAVID BOWMAN and SEPPO I. PENTTILA

Oak Ridge National Laboratory
Oak Ridge, TN 37831, USA

Performance of a neutron trap for cleaning quasi-trapped neutrons depends on what fraction of the neutron orbits are chaotic. In this paper we argue that the Lyapunov characteristic exponent is a good measure the chaos because regular orbits have Lyapunov exponent zero and chaotic orbits of a given energy have a common non-zero Lyapunov exponent. The Lyapunov exponent describes the rate of exponential divergence for infinitesimally perturbed initial conditions [1,2]. We show how to calculate the fraction of chaotic trajectories using Benettin's algorithm [1]. We evaluate the fraction of non-chaotic orbits for a trap that consists of a vertical multipole, gravity, and a current loop at the bottom of the trap.

Gravo-Magneto (GM) neutron traps offer advantages over mechanical traps for the measurement of the neutron lifetime. In a GM trap, the neutrons don't interact with matter after the trap is closed. The GM trap is filled by lowering the trapping field, allowing mechanically trapped neutrons to enter the trapping region, and then raising the field to trap the neutrons. After the field is raised, the neutrons interact only with gravity and the magnetic field. Their motion can be calculated exactly.

Neutron trajectories are either regular or chaotic. We demonstrate a method to calculate the fraction of the neutron trajectories that are chaotic. A trajectory is chaotic if when the initial point, $x0$, is displaced to $x0+dx$, the average separation between the two trajectories increases exponentially. For a given energy, the rate of increase of the logarithm of the separation is the Lyapunov exponent [1,2]. Quasi-periodic (regular) orbits have Lyapunov exponent zero.

Quasi-bound or nearly quasi-periodic orbits present a problem for measurements of the neutron lifetime. After the trap is closed, quasi-bound orbits whose energies exceed the trapping potential may remain in the trap for an arbitrarily long times. The slow escape of these quasi-bound neutrons may shift the measured neutron lifetime to lower values. Quasi-bound orbits are removed, cleaned, by inserting and then removing an absorbing cleaning surface into the trap. Chaotic trajectories are not quasi-bound because they come arbitrarily close to every point in the energetically allowed phase space as time increases and will always reach the cleaning surface. Second, numerical studies show that the angle of incidence is isotropic. Third, the probability of detection of a neutron decay product (electron or proton) depends on the location of the neutron decay. Chaotic neutron trajectories are ergodic in the sense that the time average of the detection probability is the spatial integral of some function of the configuration space (x, y, z).

As a case study, we investigated a neutron trap formed by a vertical magnetic multipole, a current loop at the bottom of the trap, and gravity. The angle of the normal to the loop can be adjusted with respect to the vertical to make the orbits to more chaotic. If the normal to the loop is vertical, horizontal and vertical modes mix slowly. As the normal is tilted away from the vertical., horizontal and vertical modes mix more rapidly. We show how to calculate the fraction of the orbits that are chaotic. We show that as the energy of neutrons E_n approaches the maximum binding energy of the trap DU, the fraction of orbits that are quasi-periotic (non-chaotic) is less than 10^{-3} with the loop tilted to 30 degrees. The geometry of our trap is shown in Figure 1.

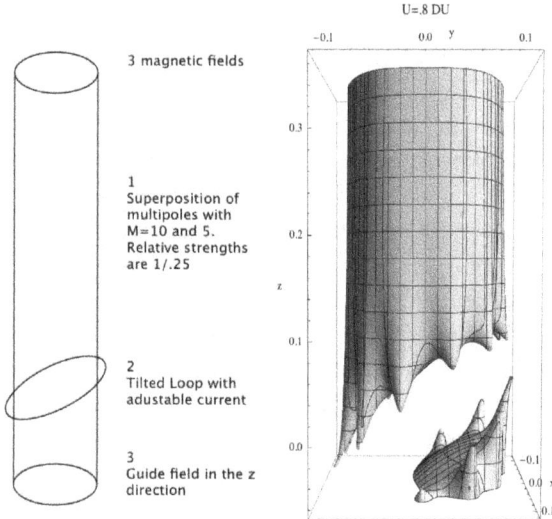

Figure 1. The principle geometry of the trap is shown in the left. At some energy U_{max}, the upper and lower equipotential surfaces coalesce and neutrons are not trapped. On the right is the equipotential surface for neutrons with $E_n = 0.8\ DU$. Only the lower quarter of the equipotential surface is shown. Less than 10^{-3} of the orbits with $E_n < 0.8\ DU$ are regular with the loop tilted to 30 degrees (Color Plate 5).

We found that important considerations in simulating neutron trajectories are: 1) Neutrons move in regular as well as chaotic orbits. 2) It is not practical to evaluate 10^9 trajectories required to model the trap with a statistical uncertainty of less than 10^{-4} for quantities that converge as $1/\sqrt{n}$. 3) It is not possible to investigate chaotic behavior of orbits using interpolated magnetic fields because the interpolation noise causes all trajectories to appear chaotic. 4) It is essential to use a symplectic integrator that preserves the phase space volume and therefore produces trajectories that exactly (to within round-off error) obey Liouville's theorem. 5) It is possible to obtain results for some problems such as cleaning and estimation the fraction of regular to chaotic trajectories by calculating quantities that converge as $1/n$ and require only few x 10^4 trajectories to obtain results accurate to 10^{-4}.

We use the 4th order symplectic integrator developed by McLachlan and Atela [3] to calculate trajectories. This integrator has no attractors and has an accuracy of less than 0.01 μm in distance and less than10^{-8} in fractional energy. In the calculations a trajectory is defined to be chaotic if the distance to any nearby trajectory increases exponentially with time. The rate of increase of the logarithm of the separation is the Lyapunov characteristic exponent (LCE) of the trajectory. The largest Lyapunov exponents for a few thousand trajectories for different initial conditions were calculated using Benettin's algorithm [4]. This algorithm tracks a reference trajectory and a nearby trajectory initially separated in phase space by a small distance, $\delta 0$, for a time, Δt. The new separation, $\delta 1$ and $\ln(\delta 1/\delta 0)$ are calculated. Then a new starting point is chosen for the nearby trajectory that is a distance $\delta 0$ from the reference trajectory on the line connecting the two trajectory endpoints. Both trajectories are again evolved for Δt, and a new separation, $\delta 2$ is calculated. This process is continued for n steps. The finite-time Lyapunov exponent, λ, is the average of $\ln(\delta n/\delta 0)$. The distribution of finite-time Lyapunov exponents has been shown to be Gaussian with a width squared that decreases inversely with n. The calculation is carried out for enough time steps to separate the quasi-periodic peak at zero from the Gaussian peak centered at the Lyapunov exponent.

The calculations were performed on a desktop computer using Mathematica 8. It was possible to calculate a few thousand trajectories in a day. The results indicate that the orbits in the trap are very chaotic for energies above $E_n > 0.1$ DU. The fraction of regular trajectories as a function of E_n/DU and the distribution of their finite time Lyapunov exponents for $E_n > 0.05$ DU are shown in Figure 2.

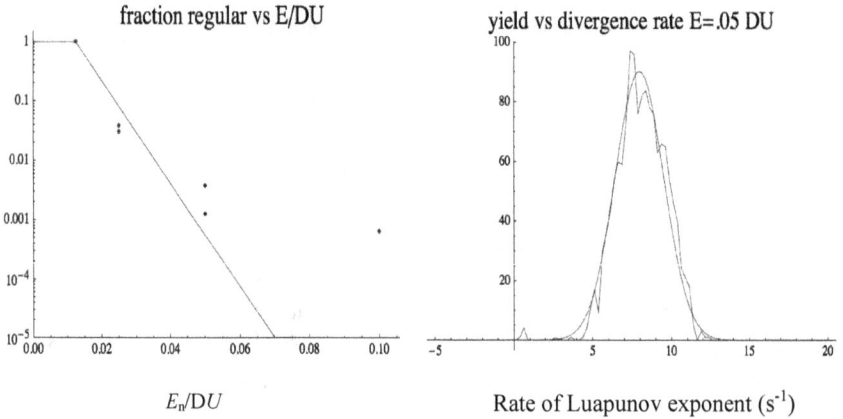

Figure 2. On the left is shown the fraction of regular trajectories as a function of the ratio of the neutron energy to the maximum trap energy, E_n/U_{max}. The fraction at $E_n/U_{max} =0.1$ is an upper limit for the regular orbits. This fraction decreases rapidly with E_n and is less than 10^{-3} for $E_n>0.05U_{max}$. In the right is a histogram of finite-time Lyapunov exponents at $E_n=0.05U_{max}$ when a few thousand trajectories were calculated. The chaotic trajectories appear in the Gaussian peak centered at 8 Hz, the small peak at 0 Hz represents one regular trajectory. The rate of divergence of the trajectories is comparable to the rate at which neutrons strike the vertical walls of the trap.

The chaotic trajectories approach arbitrarily close to every point on the energetically accessible 5-dimensional surface in phase space. Figure 2 shows that the higher-energy trajectories are chaotic with a measure of larger than 0.999. The trap can be cleaned of trajectories with energies greater than some threshold energy that is smaller than U_{max} by inserting an absorber that intersects the threshold equilibrium energy surface. For small total neutron energies, the potential energy is approximately harmonic and the trajectories are all regular. Figure 2 shows that all trajectories are regular for $E_n/U_{max}< 0.02$.

We have shown that the chaotic behavior of a model gravo-magneto trap can be quantified by calculating the Lyapunov exponents of the trajectories. For the model trap, the trajectories are all regular for small energies. As the energy increases, the fraction of regular trajectories decreases. At $E_n/U_{max}< 0.1$, the fraction of regular trajectories is small, less than 0.001. These methods can be applied to any trap that has an analytic potential.

References

1. G. Benettin, L. Galgani, J.-M. Strelcyn, Phys. Rev. A14, 2338 (1976). Informal discussions of chaos in Hamiltonian systems. Benettin's algorithm can be found in two graduate level books, "Chaos in Dynamical Systems",

E. Ott, Cambridge University Press, 2002 and "Regular and Chaotic Dynamics", A. Lichtenberg and M. Lieberman, Springer-Verlag, 1992.

2.   Ya. Pesin, Russian Math. Surveys 32, 55 (1977).

3.   R.I. McLachlan and P. Atela, Nonlinearity 5, 541 (1992).

4.   V. Baran, M. Colonna, M. Di Toro and V. Greco, Phys. Rev. Lett. 86, 4492 (2001).

Stochastic Modeling and Simulation of Marginally Trapped Neutrons

K. J. COAKLEY

National Institute of Standards and Technology
Boulder, CO 80305, USA

For a magnetic trapping experiment, I present an efficient method for simulating experimental β−decay rates that accounts for loss of marginally trapped neutrons due to wall collisions and other possible loss mechanisms. Monte Carlo estimates of time-dependent survival probability functions for the wall loss mechanism are based on computer intensive tracking of marginally trapped neutrons with a symplectic integration method and a physical model for the loss probability of a neutron when it collides with a trap boundary. The simulation is highly efficient because after all relevant survival probabilities are determined, observed neutron decay events are quickly simulated by sampling from probability distribution functions associated with each survival probability function of interest. That is, computer intensive and time-consuming numerical simulation of a large number of additional neutron trajectories is not necessary.

Keywords: Magnetic trapping; marginally trapped neutrons; monte carlo simulation; neutron lifetime; stochastic modeling; symplectic integration

1. Introduction

According to the Particle Data Group analysis of various experiments, the current best estimate of the mean lifetime of the neutron is (880.1 ± 1.1) s [1]. Systematic sources of variability are the major roadblock for reducing the overall uncertainty of the mean lifetime of the neutron [2]. In a variety of neutron lifetime experiments, marginally trapped neutrons that can escape a trapping volume before they β−decay can contribute systematic error to neutron lifetime measurements [3-7]. Here, I focus on a magnetic trapping experiment at NIST where an Ultra Cold Neutron (UCN) is produced when a 12 K neutron is scattered to near rest in liquid helium by emitting a single phonon [8]. An UCN with sufficient energy to escape the trap can be lost due to interactions with materials at the boundary of the trap before it

$\beta-$decays [9,10,11]. I present stochastic models for the survival probability for an ensemble of UCNs that account for various loss mechanisms including $\beta-$decay and loss of marginally trapped neutrons. Based on estimated survival probability functions, I present an efficient Monte Carlo method for simulating observed $\beta-$decay data. The main application of this method is to quantify systematic error in the measured neutron lifetime due to marginally trapped neutrons.

2. UCN Trajectory Model

In the NIST experiment, an UCN with sufficiently low energy is trapped in the potential field V produced by the interaction of the magnetic moment of a neutron and a spatially varying magnetic field (Figure 1) :

$$V(\mathbf{x}) = \mu|\mathbf{B}(\mathbf{x})|, \tag{1}$$

where $\mathbf{B}$ is the magnetic field, μ is the magnetic moment of the neutron and $\mathbf{x}$ is spatial location of the neutron in the Cartesian coordinates shown in Figure 1. For the NIST experiment, the gravitational force on any neutron is in the $-y$ direction. In the simulation studies presented here, the overall potential of interest is the sum of the magnetic trapping potential (Eq. 1) and the gravitational potential of the neutron $V_{grav} = m_n g y$ where g is the acceleration of gravity and m_n is the neutron mass. Henceforth, V denotes the sum of the magnetic trapping potential (Eq. 1) and V_{grav}.

In this study, I assume that UCNs are produced in a nominal cylindrical trapping volume defined by $-z_o \leq z \leq z_o$ where z is the axial coordinate and $z_o = 37.5$ cm (see Figure 1) where the potential field is static. Within the trapping volume, the distribution of the initial position of an UCN is based on an assumed beam profile for the neutrons that produce UCNs. An UCN with total energy (kinetic plus potential) greater than the minimum of the potential V_{min} on the boundary of the nominal trapping volume is defined to be marginally trapped (or above threshold). In the study presented later in this note, $V_{min} \approx 160$ neV. Neutrons with total energy below V_{min} are defined to be below threshold neutrons.

I assume that the creation time of an UCN during the loading stage is uniformly distributed and that for low energies of interest, the initial velocity of an UCN has a quadratic form [12]. That is, $f(|v|) \propto |v|^2$ To determine a neutron trajectory based on its initial position and momentum, I solve the classical equations of motion

$$\dot{\mathbf{p}} = F(\mathbf{x}) = -\nabla\mathbf{V}(\mathbf{x}), \tag{2}$$

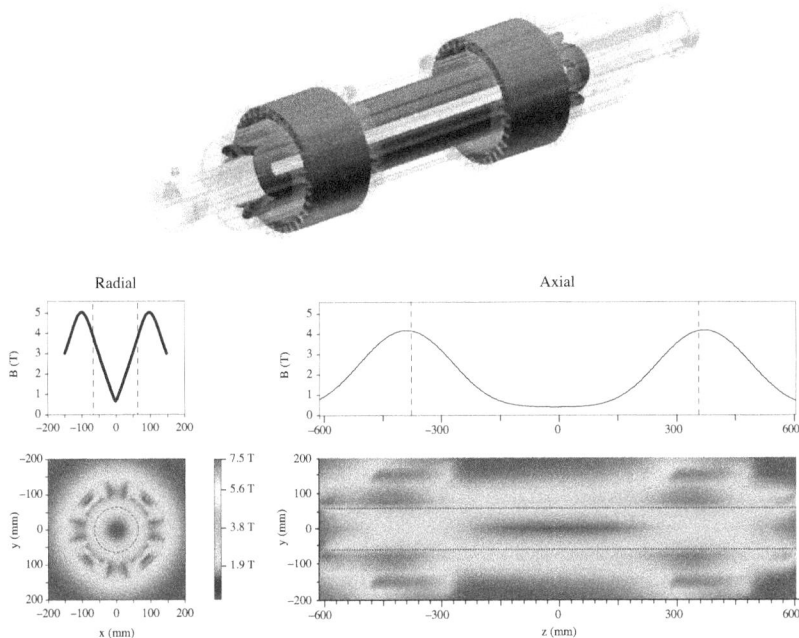

Fig. 1.   Schematic of NIST magnetic trap (Color Plate 5).

and

$$\dot{\mathbf{x}} = \frac{\mathbf{p}}{m_n}, \tag{3}$$

with an optimal fourth order symplectic integration scheme [12,13,14]. I predict $|\mathbf{B}|$ at arbitrary points in the trapping volume with a three dimensional tensor-product spline interpolant [15] where the order of the spline is four in each direction. I estimate the tensor-product B-spline coefficients from values of $|\mathbf{B}|$ computed on a grid by a numerical code which solves the Biot-Savart law numerically corresponding to the geometry of the solenoid and current bars that produce the magnetic field. The values of the potential and its gradient are evaluated at arbitrary locations given these coefficients.

When an above threshold UCN collides with the cylindrical wall or endcap boundaries, I assign it a loss probability p_{loss} according to a model based on assumed material properties [9,10,11]. After k collisions, the empirical survival probability of the UCN is

$$p_{surv}(k) = \prod_{i=1}^{k}(\ 1 - p_{loss}(i)\ ). \tag{4}$$

I track each above threshold UCN until its empirical survival probability drops below 10^{-9}. Due to chaotic scattering effects, the symplectic integration prediction for a trajectory of an UCN (Figure 2) does not converge in general as the time step parameter in the integration code is reduced [12]. Here, I assume that the mean survival probability at any time t for an ensemble of UCNs does not depend on the time step parameter even though the predicted survival probability of a particular UCN may depend on the time step parameter.

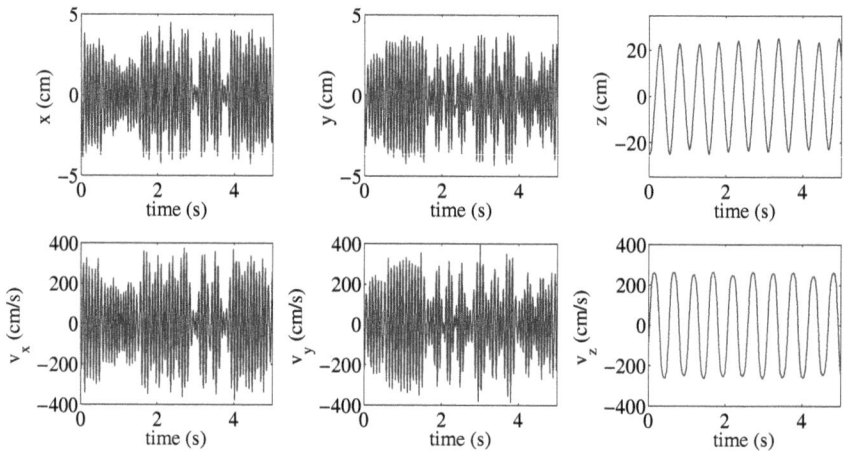

Fig. 2.   Sample trajectory determined by symplectic integration method (Color Plate 6).

3. Stochastic Model for Marginally Trapped Neutrons

Below, I present a survival analysis [16] model for marginally trapped neutrons that can be lost due to multiple mechanisms.

3.1. *Survival Probability*

For any particular loss mechanism for a particle, one can define a survival probability function $S(t)$ for a particle created at $t = 0$. For instance, for neutron β−decay,

$$S_\beta(t) = \exp(-\frac{t}{\tau_n}), \tag{5}$$

where τ_n is the mean lifetime of the neutron. In the simulation studies presented later in this work, I set τ_n to 880.1 s which, as stated earlier, is the current best estimate of the neutron lifetime. In general, for any loss mechanism, one can define a time-dependent function $\tau(t)$ as follows:

$$\frac{\dot{S}(t)}{S(t)} = -\frac{1}{\tau(t)}, \tag{6}$$

where $\dot{S}(t)$ is the time derivative of $S(t)$. Direct integration yields

$$S(t) = \exp(-\int_{s=0}^{t} \frac{ds}{\tau(s)}). \tag{7}$$

Given a survival probability function $S(t)$ for any mechanism, the loss time for that mechanism is a random variable with a cumulative probability distribution function $F(t) = 1 - S(t)$.

In the NIST experiment, UCNs can be lost due to various mechanisms including: β−decay, wall loss of marginally trapped UCNs, up-scattering and absorption by impurities such as ^{3}He. I define the survival probability associated with the loss of marginally trapped neutrons to be $S_M(t)$. Assuming that all loss mechanisms are mutually independent, the overall survival probability of an UCN created at time $t = 0$ can be expressed as

$$S(t) = S_\beta(t) S_M(t) S_{other}(t), \tag{8}$$

where $S_{other}(t)$ is the product of survival probabilities associated with each of all other loss mechanisms (excluding β−decay and marginally trapping). Since higher energy marginally trapped neutrons will tend to escape faster, one expects that the value of $\tau(t)$ associated with the loss of marginally trapped neutrons will vary with time. Hence, in general, $S_M(t)$ is not expected to be an exponential function of time.

In the NIST experiment, UCNs are created at random times during a loading stage that lasts from $t = 0$ until $t = t_L$ where $t_L = 2500$ s. After about 200 to 300 s after the trap is loaded, observations are acquired until 2800 s after the trap is loaded. Given that the production rates of below and above threshold UCNs are λ_- and λ_+ respectively, the expected number of above and below threshold UCNs at time t_L are

$$< N_+(t_L) > = \lambda_+ \int_{s=0}^{t_L} S_{other,>}(t_L - s) S_M(t_L - s) \exp(-\frac{t_L - s}{\tau_n}) ds, \tag{9}$$

and

$$< N_-(t_L) > = \lambda_- \int_{s=0}^{t_L} S_{other,<}(t_L - s) \exp(-\frac{t_L - s}{\tau_n}) ds, \tag{10}$$

where $S_{other,<}(t)$ and $S_{other,>}(t)$ are survival probability functions associated with "other" loss mechanisms for below and above threshold neutrons. If $S_{other} = 1$ for all neutrons,

$$< N_-(t_L) >= \lambda_- \tau_n (1 - \exp(\frac{-t_L}{\tau_n})). \tag{11}$$

3.2. Conditional Survival Probability

If an above threshold UCN survives the loading stage and exists at $t = t_L$, its survival probability for $t > t_L$ is

$$S^+(t) = S_M^+(t) S_{other,>}^+(t) S_\beta(t - t_L), \tag{12}$$

where

$$S_M^+(t) = c_1 \int_{s=0}^{t_L} S_{other,>}(t_L - s) S_M(t - s) \exp(-\frac{t_L - s}{\tau_n}) ds, \tag{13}$$

and

$$S_{other,>}^+(t) = c_2 \int_{s=0}^{t_L} S_{other,>}(t - s) S_M(t_L - s) \exp(-\frac{t_L - s}{\tau_n}) ds, \tag{14}$$

where c_1 and c_2 are integration constants that ensure that $S_M^+(t_L) = S_{other,>}^+(t_L) = 1$. The conditional survival probability function of a below threshold UCN, for $t > t_L$, is

$$S^-(t) = S_{other,<}^-(t) S_\beta(t - t_L), \tag{15}$$

where

$$S_{other,<}^-(t) = c_3 \int_{s=0}^{t_L} S_{other,<}(t - s) \exp(-\frac{t_L - s}{\tau_n}) ds, \tag{16}$$

where c_3 ensures that $S_{other,<}^-(t_L) = 1$. To illustrate this method, for the special case where for all t $S_{other,>}(t) = S_{other,<}(t) = 1$, I show estimates of $S_M(t)$ and $S_M^+(t)$ for an ensemble of above threshold UCNs for a particular wall loss probability model (Figure 3).

3.3. Neutron Decay Signal

The predicted β−decay signal after loading the trap for times $t_L \leq t \leq t_{end}$ is:

$$r_\beta(t) = \frac{< N_-(t_L) > S^-(t) + < N_+(t_L) > S^+(t)}{\tau_n}. \tag{17}$$

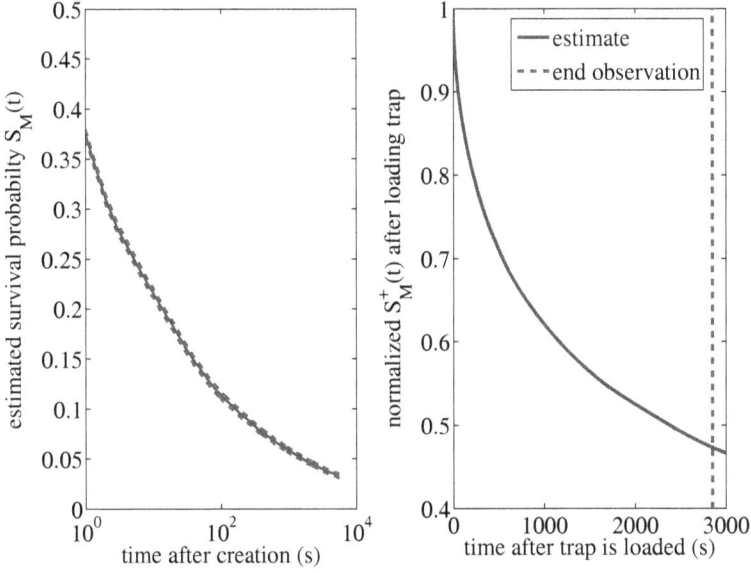

Fig. 3. Left: Monte Carlo estimate of the survival probability associated with the wall loss mechanism for an ensemble of above threshold UCNs and associated $\pm$-1 standard error bands. Right: conditional survival probability for wall loss mechanism for UCNs that survive until loading stage is completed. In this simulation study, UCNs are lost by either β-decay or the marginally trapped loss mechanism. In this simulation study $\tau_n = 880.1$ s (Color Plate 6).

For the special case where $S_{other}(t) = 1$ for all t for both below and above threshold UCNs,

$$r_\beta(t) = f_{exp}(t) + f_c(t), \tag{18}$$

where

$$f_{exp}(t) = \frac{<N_-(t_L)> + <N_+(t_L)> S_M^+(t_{end})}{\tau_n} \exp(-\frac{t - t_L}{\tau_n}), \tag{19}$$

and

$$f_c(t) = \frac{<N_+(t_L)> (\ S_M^+(t) - S_M^+(t_{end})\ )}{\tau_n} \exp(-\frac{t - t_L}{\tau_n}). \tag{20}$$

Given Eqns. 19 and 20, one can simulate β-decay data by sampling from a mixture of an exponential and an non-exponential probability density function (Figure 4). For the non-exponential part of the mixture, f_c, the conditional cumulative probability density function for loss times due to wall collisions during the interval (t_L, t_{end}) is

$$F_M(t) = 1 - \frac{(\ S_M^+(t)\ -\ S_M^+(t_{end})\ )}{(\ S_M^+(t_L)\ -\ S_M^+(t_{end})\ )}. \tag{21}$$

At $t = t_L$, the ratio of the non-exponential decay rate to the exponential decay rate can be regarded as a contamination ratio r_c where

$$r_c = \frac{< N_+(t_L) > (\ S_M^+(t_L)\ -\ S_M^+(t_{end})\ )}{< N_-(t_L) > + < N_+(t_L) > S_M^+(t_{end})}. \tag{22}$$

The quantities $< N_+(t_L) >$, $< N_-(t_L) >$, and $S_M^+(t)$ would be estimated by Monte Carlo methods (see Figure 3).

For the NIST magnetic trapping experiment, the neutron lifetime is estimated by fitting exponential models to background-corrected neutron β−decay rate data. To quantify systematic error due to neglecting marginally trapped UCNs, the same models (which neglect marginally trapped UCNs) are fit to simulated neutron β−decay data such as shown in Figure 4. As a caveat, if one had very high confidence in the Monte Carlo estimates of $S_M^+(t)$ and $S_M^-(t)$, one might fit a model based on Eq. 18 (or an extended version of it to account for background-correction) to observed data. That is, $S_M^+(t)$ and $S_M^-(t)$ might be included in the prediction model fit to observed data. The alternative approach is to ramp the trapping potential down and then back up again so as to purge marginally trapped UCNs. The stochastic model and proposed Monte Carlo simulation method presented here is useful for quantifying how well such ramping strategies purge above threshold UCNs for particular wall loss probability models and field ramping strategies. For more discussion of purging, see [4,17].

In general, for the case where $S^{+/-}(t) = S_\beta(t - t_L) \prod_{i=1}^{k} S_i^{+/-}(t - t_L)$ one would simulate a β−decay loss time t_β and loss times for each of the other k mechanisms for either below threshold or above threshold UCNs. For each subsample, a necessary condition for observing the decay of a particular neutron is that $t_\beta < \min(t_1, t_2, \cdots, t_k)$.

For the NIST magnetic trapping experiment, preliminary Monte Carlo simulation studies indicate that the marginally trapped neutron loss mechanism produces a systematic error in the neutron lifetime measurement of approximately -40 s for a typical experiment where the trapping potential is static. In this study, the true neutron lifetime was set to 880.1 s and the only loss mechanisms considered were due to β−decay and the interaction of marginally trapped neutrons with materials at the boundary of the trap. Preliminary Monte Carlo simulation experiments indicate that purging marginally trapped neutrons by magnetic field ramping can reduce this systematic error to very low levels at the expense of increasing the statistical uncertainty of the measurement.

In the NIST experiment, detection efficiency of β−decay varies with axial location. Accounting for this spatial dependence, or quantifying systematic error associated with neglecting it, is an ongoing research topic.

The Monte Carlo estimates of survival probabilities are uncertain due to sampling variability. Further, uncertainties in the assumed value τ_n and the assumed wall loss probability model will produce uncertainties in these survival probabilities. Quantifying the overall uncertainty of the estimated survival probability functions due to uncertainties in the assumed wall loss probability model and the assumed value of τ_n, and other sources of variability is an ongoing research topic.

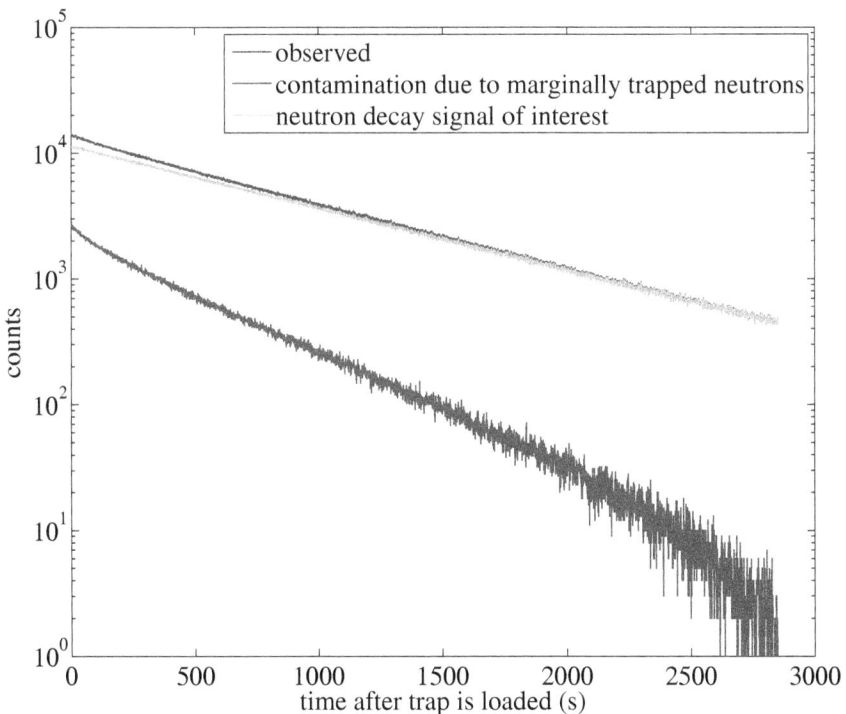

Fig. 4. Simulated β−decay data. Neutron signal of interest corresponds to Eq. 19. The contamination signal corresponds to Eq. 20. In this simulation study $\tau_n = 880.1$ s (Color Plate 7).

Acknowlegements

I thank Paul Huffman, Pieter Mumm, Craig Huffer, Susan Seestrom, Grace Yang and an anonymous reviewer for useful comments. Contributions by staff of NIST, an agency of the US government, are not subject to copyright in the US.

References

1. J. Beringer et al. (Particle Data Group), PR D86, 010001 (2012) (URL: http://pdg.lbl.gov).
2. F.E. Wietfeldt and G.L. Greene, Reviews of Modern Physics 8383 1173 (2011).
3. S. Paul, Nucl. Instr. and Methods A611 (2009) 157.
4. P. R. Huffman, C. R. Brome, J. S. Butterworth, K. J. Coakley, M. S. Dewey, S. N. Dzhosyuk, R. Golub, G. L. Greene, K. Habicht, S. K. Lamoreaux, C. E. H. Mattoni, D. N. McKinsey, F. E. Wietfeldt, J. M. Doyle Nature 403, 62-64 (6 January 2000).
5. R. Picker, I. Altarev, J. Bröcker, E. Gutsmiedl, J. Hartmann, A. Müller, S. Paul, W. Schott, U. Trinks, and O. Zimmer, J. Res. Natl. Inst. Stand. Technol. 110, 357-360 (2005).
6. K. Leung and O Zimmer, Nucl. Instr. and Methods A611 (2009) 181.
7. P.L. Walstrom, J.D. Bowman, S.I. Penttila, C. Morris, A. Saunders, Nucl. Instr. and Methods A599 (2009) 82.
8. R. Golub and J.M. Pendlebury, Phys. Lett. 53A, 133 (1975).
9. R. Golub, D. Richardson, S.K. Lamoreaux, Ultra Cold Neutrons, Taylor and Francis (1991).
10. A. Steyerl, S. S. Malik, A. M. Desai, and C. Kaufman, Phys. Rev. C 81, 055505 (2010).
11. C.M. Oshaughnessy, PhD thesis, North Carolina State University (2010).
12. K. J. Coakley, J. M. Doyle, S. N. Dzhosyuk, L. Yang, P. R. Huffman, J. Res. Natl. Inst. Stand. Technol. 110, 367-376 (2005).
13. R. I. McLachlan and P. Atela, Nonlinearity 5, 541-562 (1992).
14. P.J. Channel and C. Scovel, Nonlinearity 3, 231 (1990).
15. C. de Boor, A practical guide to splines, Springer-Verlag, New York (1978).
16. D.R. Cox and D. Oakes, Analysis of Survival Data, Monographs on Statistics and Applied Probability 21, Chapman and Hall/CRC, Boca Raton (1998).
17. L. Yang, PhD thesis, Harvard University (2006).

Spin Flip Loss in Magnetic Storage of Ultracold Neutrons

A. STEYERL*, C. KAUFMAN, G. MÜLLER, S. S. MALIK, and A. M. DESAI

Department of Physics, University of Rhode Island,
Kingston, Rhode Island 02881, USA
* asteyerl@mail.uri.edu
www.phys.uri.edu

We analyze the depolarization of ultracold neutrons confined in a magnetic field configuration similar to those used in existing or proposed magneto-gravitational storage experiments aiming at a precise measurement of the neutron lifetime. We use an approximate quantum mechanical analysis such as pioneered by Walstrom *et al.* [Nucl. Instrum. Methods Phys. Res. A **599**, 82 (2009)]. Our analysis is not restricted to purely vertical modes of neutron motion. The lateral motion is shown to cause the predominant depolarization loss in a magnetic storage trap.

1. Introduction

The neutron lifetime τ_n is an important parameter in tests of the Standard Model of particle physics. It also affects the rate of helium production in the early universe and the energy production in the sun. The current Particle Data Group (PDG) average is $\tau_n = 880.1 \pm 1.1$ s.[1] However, the value of one experiment,[2] which reported the lowest measurement uncertainty, namely ~ 0.8 s, is located some 3.5 s below the bulk of other data in the PDG collection.[3–8] The latter are consistently grouped around 882.0 s ($\pm$ 1.0 s).[9] As a possible way of advancing this field, storage of polarized ultracold neutrons (UCNs) in a magnetic trap has been pioneered by Paul *et al.*[10] and is currently being pursued vigorously by several groups worldwide.[11–15] In magnetic traps there are no wall losses, the slow loss due to quasi-stable orbits is serious but believed to be manageable by avoiding regular orbits,[14] and the potential loss due to depolarization, defined as spin flip relative to the local field direction, is argued to be negligible.

Until recently UCN depolarization estimates[16,17] have been based on Majorana's quasi-classical result of 1932[18] for a free polarized particle with magnetic moment moving with constant velocity vector through a non-uniform static magnetic field of specific form. For magnetic field parameters as currently used or proposed for UCN storage, the spin-flip probability D for passage through the field would be of order $\exp(-10^6)$, thus immeasurably small. Recently, Walstrom et al.[14] pointed out that the values of D for confined, rather than freely moving, neutrons are much larger. For a UCN moving along a vertical path in the storage system proposed by them, D was estimated to be in the range $D \sim 10^{-20}$ to 10^{-23}. This is much larger than the Majorana value but still negligible in any actual or projected neutron lifetime experiment.

Using a simplified model of magnetic field distribution we extend that theory to include arbitrary UCN motion with both vertical and horizontal velocity components, confined to the vertical space between upper and lower turning points that depend only on the UCN energy for vertical motion. In our model (introduced in Sec. 2) the magnetic field magnitude B is uniform within any horizontal plane, so there is no horizontal component of magnetic force. Therefore the neutron moves with constant velocity in the horizontal z- and x-directions. We show that D could reach a level approaching the tolerance limit for a high-precision neutron lifetime measurement unless precautions are taken.

Our model field is close to the "bathtub configuration" of Ref. 14 but the lateral confinement of UCNs, achieved there by double curvature of the magnetic mirror, is simulated differently. The magnetic mirror is horizontal and extends to infinity in both lateral dimensions. However, one could imagine the presence of ideal vertical mirrors reflecting the UCNs back and forth in the horizontal directions without any change in the analysis. More specifically, we use an infinite ideal planar Halbach array,[19] which is free of the field ripples present in actual realizations.[14]

The topic of UCN depolarization in magnetic storage or in mirror reflection in a magnetic field raises interesting questions of quantum interpretation. We postulate that the depolarization rate expected for a UCN magnetic storage experiment is determined by the current of UCNs in the "wrong" spin state. Neutrons in this (high-field seeking) state exit the system at the lower and upper turning points, whereas neutrons in the "correct" (high-field repelled) state are reflected and return to the storage space. Exiting neutrons could be counted by detectors placed just outside the turning points. In the Copenhagen interpretation, such a measurement (actual or hypothetical) resets the UCN wave function to a pure state of high-field repelled neutrons. The spin state then evolves as described by the spin-dependent Schrödinger equation (or its semi-classical analog) until the next "measurement" takes place and the reset is repeated. A more comprehensive report of the present work can be found in Ref. 22, where we have also analyzed UCN depolarization in reflection from a nonmagnetic mirror placed into a nonuniform magnetic field.

2. Magnetic Field Distribution

As illustrated in Fig. 1, an ideal Halbach array[19] of permanent magnets of thickness d covering the (zx)-plane generates a magnetic field

$$\mathbf{B}_H(x,y) = B_0 e^{-Ky}(\hat{\mathbf{x}}\cos Kx - \hat{\mathbf{y}}\sin Kx), \tag{1}$$

where $B_0 = B_{\mathrm{rem}}(1 - e^{-Kd})$ is determined by the remanent field B_{rem}. We choose similar parameters as in Ref. 14: $L = 2\pi/K = 5.2$cm, $d = 2.54$cm and $B_0 = 0.82$T.

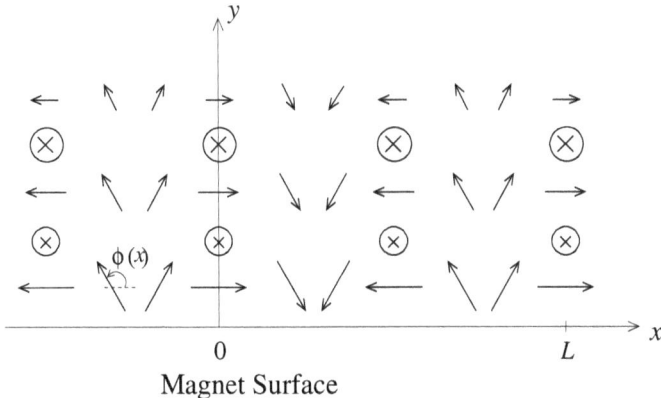

Fig. 1. The arrows show the Halbach magnetic field $\mathbf{B}_H$ as it rotates in the (xy)-plane. Its magnitude B_H decreases exponentially with height y and is represented by the arrow length using a log scale. The stabilization field $\mathbf{B}_1$ in the z-direction increases slowly with y as symbolized by the crosses of variable size.

In the actual scheme,[14] the uniform rotation is replaced by dividing the rotation period L into four blocks, each of length $L/4$ and with the same magnetization M, but with an angle of 90° between the directions of $\mathbf{M}$ in adjacent blocks (schematically represented as ... ←↓→↑← ...). Alternative designs are in the form of vertical or horizontal cylinders.[11–13,15]

For the stabilization field we use $\mathbf{B}_1 = \hat{z}B_{10}\rho/(\rho-y)$ with $\rho = 1.5$ m. In Ref. 14 a value of 0.05 to 0.1 T was proposed for B_{10}. We consider similar field strengths down to the mT range.

3. Basic Equations

The wave function for a UCN moving in the magneto-gravitational field of the trap is a linear superposition of the two eigenstates of the magnetic moment interaction Hamiltonian,

$$\mathcal{H}_m = -\mu_n \sigma \cdot \mathbf{B}, \tag{2}$$

where $\mu_n = -1.913\mu_N$ is the neutron magnetic moment in terms of the nuclear magneton $\mu_N = 0.505 \times 10^{-26}$ J/T, σ is the Pauli spin operator, and $\mathbf{B}$ is the local magnetic field. The two eigenstates χ^+ and χ^- of $\mathcal{H}_m$ satisfy the eigenvalue equations,

$$\mathcal{H}_m\chi^\pm = \pm|\mu_n|B\chi^\pm, \tag{3}$$

with spin parallel and antiparallel to $\mathbf{B}$, respectively. These spin eigenfunctions are obtained by spin rotation from the z-axis to the direction of $\mathbf{B}$ through angles θ and ϕ. The polar angle is $\theta = \cos^{-1}(B_z/B) = \sin^{-1}(B_{xy}/B)$, where $B_z = B_1$ is due to the bias field $\mathbf{B}_1$ and $B_{xy} = B_H$ is the magnitude of the Halbach field $\mathbf{B}_H$. The azimuthal angle in the (xy)-plane is $\phi = \sin^{-1}(B_y/B_{xy}) = \tan^{-1}(B_y/B_x)$.

For the Halbach field configuration (1) we have $\phi = -Kx$. Thus ϕ depends only on x, while θ depends only on y. Performing the spin rotation through angles θ and ϕ we obtain for the spin basis vectors with quantization axis along $\mathbf{B}$,

$$\chi^- = \begin{pmatrix} e_-s \\ -c \end{pmatrix}, \quad \chi^+ = \begin{pmatrix} c \\ e_+s \end{pmatrix}, \tag{4}$$

where $s = \sin(\theta/2)$, $c = \cos(\theta/2)$ and $e_\pm = \exp(\pm i\phi) = \exp(\mp iKx)$. We write the dependence of the wave function on position and spin in the form

$$\chi = \alpha^{(3)}(x,y,z)\chi^+ + \beta^{(3)}(x,y,z)\chi^-, \tag{5}$$

where we have used the superscript "(3)" to indicate that $\alpha^{(3)}(x,y,z)$, $\beta^{(3)}(x,y,z)$ are functions of the three space coordinates. By contrast, the functions $\alpha(y)$ and $\beta(y)$, introduced below, depend on y only. χ satisfies the eigenvalue equation,

$$E\chi = \left[-\frac{\hbar^2}{2m}\nabla^2 + mgy + |\mu_n|\sigma \cdot \mathbf{B}\right]\chi, \tag{6}$$

for a neutron of mass m with constant total energy E moving in a uniform gravitational field of magnitude g and a non-uniform magnetic field $\mathbf{B}$.

As in Ref. 14 we use the WKB approximation[23] and keep only terms that contain the derivatives of the field variables (θ and ϕ) in lowest order. Their variation is on the scale of centimeters whereas the waves in real space, $\alpha^{(3)}$ and $\beta^{(3)}$, vary on the micrometer scale, i.e. $\sim 10^4$ times faster. With the UCN initially in a pure $(+)$ spin state, we obtain

$$\nabla^2\chi = (\alpha^{(3)}_{xx} + \alpha^{(3)}_{yy} + \alpha^{(3)}_{zz})\chi^+$$
$$+ [\beta^{(3)}_{xx} + \beta^{(3)}_{yy} + \beta^{(3)}_{zz} + e^{-iKx}(-\theta_y\alpha^{(3)}_y + iK\alpha^{(3)}_x \sin\theta)]\chi^-, \quad (7)$$

for scenarios where $|\beta^{(3)}| \ll |\alpha^{(3)}|$ holds. The expressions multiplying χ^+ and χ^- can be simplified by noting that the x- and z-dependences of $\alpha^{(3)}$ have the plane-wave form $e^{ik_x x}e^{ik_z z}$ and $\beta^{(3)}$ is proportional to $e^{-iKx}e^{ik_x x}e^{ik_z z}$. The wave numbers k_x and k_z are constant and e^{-iKx} represents a Bloch-wave modulation due to the periodicity of the Halbach field. In practice, k_x and k_z are of order μm^{-1}, thus much larger than K and θ_y, both of which are of order cm^{-1}. Thus we can factor Eq. (7) in the form

$$\nabla^2\chi = e^{ik_x x}e^{ik_z z}\{[\alpha'' - (k_x^2 + k_z^2)\alpha]\chi^+$$
$$+ e^{-iKx}[\beta'' - (k_x^2 + k_z^2)\beta - (\theta'\alpha' + Kk_x\alpha \sin\theta)]\chi^-\}, \quad (8)$$

simplifying the notation. In Eq. (8) and henceforth, $\alpha(y)$ and $\beta(y)$ stand for the y-dependent parts of the wave function only, and differentiation with respect to y is denoted by primes. We also drop the subscript y. We thus write $\alpha^{(3)}(x, y, z) = \alpha(y)e^{ik_x x}e^{ik_z z}$ and $\beta^{(3)}(x, y, z) = \beta(y)e^{-iKx}e^{ik_x x}e^{ik_z z}$.

Substituting Eq. (8) into the eigenvalue equation (6) gives[14] two coupled equations, one for spinor χ^+ and the other for χ^-:

$$E\alpha = -\frac{\hbar^2}{2m}\left[\alpha'' - (k_x^2 + k_z^2)\alpha\right] + mgy\alpha + |\mu_n|B\alpha \quad (9)$$

and

$$E\beta = -\frac{\hbar^2}{2m}\left[\beta'' - (k_x^2 + k_z^2)\beta - (\theta'\alpha' + Kk_x\alpha \sin\theta)\right] + mgy\beta - |\mu_n|B\beta. \quad (10)$$

The WKB solution of (9) is[14]

$$\alpha(y) = k_+^{-1/2}(y)\exp\left(\pm i\Phi_+(y)\right), \quad (11)$$

where

$$\frac{\hbar^2 k_\pm^2(y)}{2m} = E - \frac{\hbar^2}{2m}(k_x^2 + k_z^2) + mg(y_0 - y) \mp |\mu_n|B(y). \quad (12)$$

Here y_0 is the greatest height a neutron of energy E and given k_x and k_z would reach in the gravitational field if the magnetic field were switched off, and $\Phi_+(y) = \int_{y_s}^{y} k_+(u)du$ is the phase angle for the $+$ spin state, accumulated between the start of vertical motion and the position y. The initial height y_s for motion upward is assumed to be that of the lower turning point, thus $y_{s+} = y_l$, and for motion downward the initial level is taken at the upper turning point, $y_{s-} = y_u$. The additional $+$ or $-$ sign in the argument of the exponential function in (11), in front of Φ_+, refers to this direction of the motion; $+$ for upward and $-$ for downward, as in Ref. 14.

The WKB wave function (11) is normalized to a constant particle flux $\hbar/m$ in y-direction. For the spin-flipped UCNs, the flux in the y-direction is the measure of the probability of depolarization, as shown below. At the classical turning points, where $k_+ = 0$, the WKB form (11) diverges and has to be replaced by the Airy function.[14] Matching the Airy function to the WKB approximation is described in detail in Ref. 22.

It follows from Eq. (10) that the wave function $\beta(x, y)$ for the spin flipped component is determined by the inhomogeneous second-order differential equation,

$$\beta''(y) + k_-^2(y)\beta(y) = \theta'(y)\alpha'(y) + Kk_x\alpha(y)\sin\theta(y), \qquad (13)$$

and may be written[14] in the WKB form

$$\beta(y) = k_-^{-1/2}(y)\exp\left(\pm i\Phi_-(y)\right)f(y), \qquad (14)$$

where the function $f(y)$ modulating the WKB wave represents the amplitude of spin flip. The phase accumulated since the start at a turning point, $\Phi_-(y) = \int_{y_s}^{y} k_-(u)du$, always has a larger magnitude than the phase $\Phi_+(y)$ for $\alpha(y)$ since k_- is greater than k_+ (except in zero magnetic field).

Thus the governing equation for $\beta(y)$ is the second-order differential equation

$$\beta''(y) + k_-^2(y)\beta(y) = [\pm ik_+\theta'(y) + Kk_x\sin\theta(y)]\alpha(y). \qquad (15)$$

We have carried out the differentiation of $\alpha(y)$ using the WKB rule with the result $\alpha' = \pm ik_+\alpha(y)$, where the $+$ sign applies to upward motion and the $-$ sign to downward motion. This replacement is valid except within a few μm of the turning points.

Our Eq. (15) is consistent with Eq. (28) of Ref. 14 except for the additional, k_x-dependent term on the right-hand side. It is present because we include motion with finite lateral momentum $\hbar k_x$. We will show that this new term dominates the UCN depolarization, since UCNs moving in x-direction are exposed to the strong field ripple due to the rotating Halbach field, in our model field as well as for the "bathtub system".[14]

4. Depolarization in Magnetic Storage

It has been shown in Ref. 22 that Eq. (15) can be solved by straightforward integration. For the downward motion we obtain

$$\beta(y) = k_-^{-1/2}(y)P(y)\exp\left(-i\Phi_+(y)\right), \tag{16}$$

where $P(y) = [iU(y) + V(y)]/W(y)$, $U(y) = -\sqrt{k_+ k_-}\,\theta'$, $V(y) = \sqrt{k_-/k_+}Kk_x\sin\theta$. $W(y) = k_-^2(y) - k_+^2(y) = (4m/\hbar^2)|\mu_n|B(y)$ depends only on the magnitude $B(y)$ of the local magnetic field.

The phase Φ_+ (with the index $+$) indicates that this wave for the $(-)$ spin state propagates, not with wave number k_-, but with the same wave number k_+ as the $(+)$ spin state, as it should.

Equation (16) represents a particular solution of (15) and we could add any solution $\beta_{h\pm}(y)$ of the homogeneous equation $\beta_h''(y) + k_-^2(y)\beta_h(y) = 0$. In the WKB framework, these solutions are $\beta_{h\pm}(y) = C_\pm k_-^{-1/2}(y)\exp\left(\pm i\Phi_-(y)\right)$ with arbitrary constants $C_\pm$. These functions represent a constant current in the upward (downward) direction for the $+$ $(-)$ sign. Thus the same current enters and exits the storage space, resulting in a zero contribution to the net flux out and, therefore to the depolarization.

Reverting to solution (16), we identify the net depolarization over the path from upper turning point y_u to y_l as the current of spin-flipped UCNs at the endpoint y_l. This current represents the net flux out of the storage space since no flux enters at y_u.

At an arbitrary position y along the way the current $j_-(y)$ is given by

$$j_-(y) = \frac{\hbar}{m}\text{Re}\left[i\beta^*(y)\left(\frac{d\beta}{dy}\right)\right],$$

$$= \frac{\hbar}{m}\left(\frac{k_+}{k_-}\right)|P|^2 = \frac{\hbar}{m}\frac{k_+^2\theta'^2 + K^2 k_x^2\sin^2\theta}{(k_-^2 - k_+^2)^2}. \tag{17}$$

The depolarization probability $(m/\hbar)j_-(y)$ is plotted in Fig. 2 for UCNs with energy for vertical motion determined by the "drop heights" $y_0 = 10$ cm and 45 cm, a bias magnetic field $B_{10} = 0.005$ T and $v_x = 3$ m/s. As in Ref. 14 we see a sharp peak at the y-position where θ' is large, and a decrease as the particle drops further down. The third curve in Fig. 2 is for $y_0 = 45$ cm, $B_{10} = 0.005$ T and $v_x = 0$. The peak value and the decrease on the upper side are quite similar. Below the peak position the curve for $v_x = 0$ decreases faster than for $v_x = 3$ m/s.

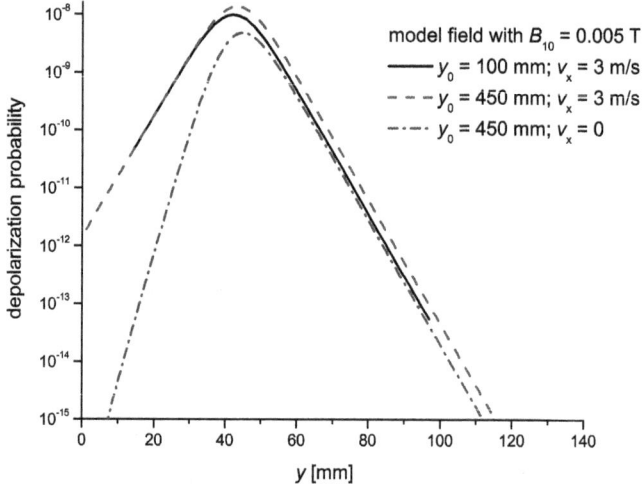

Fig. 2. Depolarization probability, given by Eq. (18) multiplied by $m/\hbar$, as a function of neutron position for drop heights $y_0 = 450$ mm and 100 mm, stabilization field parameter $B_{10} = 0.005$ T, and neutron velocity component $v_x = 3$ m/s or zero. The sharp peak occurs in the region where the gradient of field angle θ is largest.

The current leaving the storage space at $y = y_l$ is

$$j_l = \frac{\hbar}{m}\left(\frac{k_{+l}}{k_{-l}}\right)|P_l|^2 = \frac{\hbar}{m}\frac{k_{+l}^2\theta_l'^2 + K^2 k_x^2 \sin^2\theta_l}{(k_{-l}^2 - k_{+l}^2)^2} = \frac{\hbar}{m}\frac{K^2 k_x^2}{k_{-l}^4}\sin^2\theta_l, \quad (18)$$

where the index l refers to the values at $y = y_l$ and the last expression uses the fact that k_+ vanishes at the turning points.

The dependence of (18) on the field variables is established by noting that $\sin^2\theta = B_H^2/B^2$, $k_{-l}^4 \sim B_l^2$ and $K^2 k_x^2 = (m/\hbar)^2\omega^2$, where $\omega = 2\pi v_x/L$ is the frequency of the Halbach field as seen by the moving UCN.

For upward motion from y_l to y_u we get the result for the current (18) with all indices l replaced by u. The quantities relevant for the spin-flipped current leaving the system at the upper turning point are determined by the field angle θ_u and by k_{-u} at y_u. The combined depolarization loss for one reflection on the magnetic field, i.e. for one complete round trip down and up thus becomes

$$\frac{m}{\hbar}(j_l + j_u) = K^2 k_x^2 \left(\frac{\sin^2 \theta_l}{k_{-l}^4} + \frac{\sin^2 \theta_u}{k_{-u}^4} \right). \tag{19}$$

In magnetic storage the UCNs have positive and negative velocities in any direction and, for a low-energy Maxwell spectrum, with uniform probability density in phase space. Thus we replace k_x^2 by its mean value, $k_{x,max}^2/3$, for $-k_{x,\max} < k_x < +k_{x,\max}$.

As a final step we establish the explicit connection between the mean loss current and the rate of depolarization, $\tau_{\rm dep}^{-1}$, which should be negligible compared to the neutron β-decay rate in a neutron lifetime measurement. For given neutron energy for vertical motion, i.e. fixed turning heights at y_l and y_u, the depolarization rate (in s^{-1}) is determined by the loss current divided by the number of UCNs in the field-repelled spin state present in the trap,

$$N = 2 \int_{y_l}^{y_u} |\alpha(y)|^2 dy = 2 \int_{y_l}^{y_u} \frac{1}{k_+(y)} dy. \tag{20}$$

We have used $|\alpha(y)|^2$ as the density and the factor 2 takes into account that both downward and upward moving UCNs are in the trap at the same time.

Since $k_+ = (m/\hbar)v_+$ and $dy = v_+ dt$, the expression in (20) equals $(\hbar/m)T$ where T is the time required for one round trip down and up. Thus, the depolarization rate is

$$\tau_{\rm dep}^{-1} = \frac{\langle j_l + j_u \rangle}{N} = \frac{m}{\hbar} \frac{\langle j_l + j_u \rangle}{T} = K^2 \left(\frac{k_{x,\max}^2}{3} \right) \left(\frac{\sin^2 \theta_l}{k_{-l}^4} + \frac{\sin^2 \theta_u}{k_{-u}^4} \right) \frac{1}{T}. \tag{21}$$

This shows that the loss current (19) of spin-flipped UCNs equals the loss per round trip, i.e. for one bounce in the magnetic field.

For comparison with actual experiments we have to average (21) also over v_y. As a measure of v_y for a stored UCN we choose its value at the neutral plane $y = y^{(n)}$, where the gravitational force is compensated by the magnetic force pushing upward, i.e. where $|\mu_{\rm n}|(dB/dy) = -mg$. This is the plane where the UCNs with the lowest energy for vertical motion reside. In our field model, a UCN with vertical velocity $v_y^{(n)} = 0$ in the neutral plane floats or moves along the plane at constant speed. In actual confinement fields as in Ref. 14 they would follow closed or open paths on the curved neutral surface. For small values of $v_+^{(n)}$ the vertical motion is a classical harmonic oscillation with natural frequency $\omega_0 = \sqrt{dg_+/dy}$, where

$g_+ = g + (|\mu_n|/m)(dB/dy)$ is the net downward acceleration. This implies that for small oscillations about the neutral plane the time for a round trip becomes $T = 2\pi/\omega_0 = 2\pi(dg_+/dy)^{-1/2}$. For larger vertical velocities the oscillator potential is strongly anharmonic but the drop height y_0, used originally as a measure of energy for vertical motion, is unambiguously determined by $v_+^{(n)}$. Therefore, if we plot the depolarization rate (12) versus $v_+^{(n)}$, rather than y_0, the mean height of this curve in the range from $v_+^{(n)} = 0$ to its maximum value for the stored UCN spectrum directly represents the average value of depolarization rate for a Maxwell spectrum.

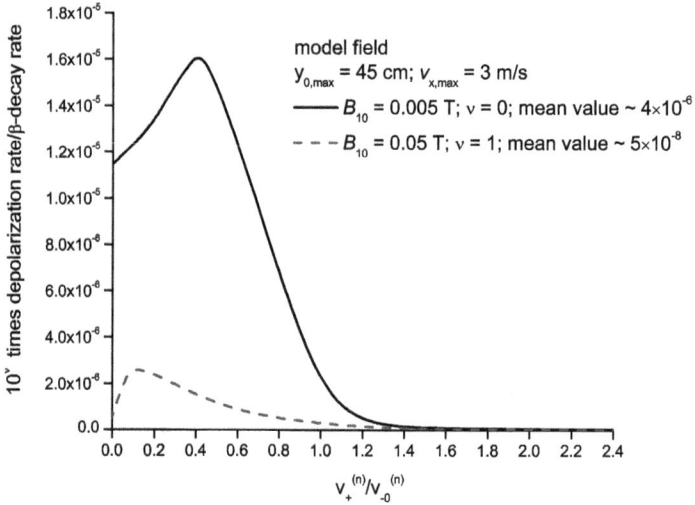

Fig. 3. Ratio between mean depolarization rate, given by Eq. (21), and neutron β-decay rate, plotted as a function of vertical velocity component $v_+^{(n)}$ in the neutral plane (where the gravitational and magnetic forces are balanced). The curve for $B_{10} = 0.005$ T is plotted to scale ($\nu = 0$) and the curve for $B_{10} = 0.05$ T is plotted with magnification factor 10^1 ($\nu = 1$). Their difference by about two orders of magnitude shows the strong suppression of depolarization by a stabilization field of sufficient strength. For a Maxwell spectrum, the mean height of the curves over the range of the abscissa, from 0 to 2.5 for $B_{10} = 0.05$T and from 0 to 4.7 for $B_{10} = 0.005$T, directly determines the average over the full spectrum.

Such a plot is presented in Fig. 3, where we have normalized $v_+^{(n)}$ to $v_{-0}^{(n)} = 2\sqrt{|\mu_n|B^{(n)}/m}$, the y-velocity for the spin-flipped state on the neutral plane for $v_+^{(n)} = 0$. For $y_{0,\max} = 45$ and $v_{x,\max} = 3$ m/s the mean depolarization rate, normalized to the β-decay rate $1/\tau_n$, is $\tau_n \langle \tau_{\text{dep}}^{-1} \rangle = 4 \times 10^{-6}$ for $B_{10} = 0.005$ T and about two orders of magnitude less for $B_{10} = 0.05$T.

The largest contribution to depolarization originates from UCNs with fairly low energy of vertical motion. They move through the field almost horizontally, with small vertical oscillations about the neutral plane. The result is plausible since these UCNs spend the largest fraction of time in the region where the field rotates rapidly in the moving reference frame.

5. Conclusion

We extend the analysis of Ref. 14 to include arbitrary UCN orbits with lateral velocity components. As a main result of the extension we find that the lateral x-component of motion in the plane of the Halbach field makes the dominant contribution to depolarization while the depolarization due to the vertical motion is insignificant. As a result, some previous estimates of depolarization probability may have been overoptimistic. For the parameters of[14] (0.05-0.1 T for B_{10}) we estimate on the basis of Fig. 3 that even a measurement of the neutron lifetime with precision 10^{-5} should be possible (disregarding other potential limitations) but the safety margin may be smaller than previously expected.

Acknowledgments

We are grateful to R. Golub for useful comments and to V. Ezhov, C. Liu, and A. Young for having drawn our attention to depolarization in magnetic UCN confinement.

References

1. K. Nakamura *et al.*, (Particle Data Group), *J. Phys. G* **37**, 075021 (2010), and 2012 partial update.
2. A. Serebrov *et al.*, *Phys. Lett. B* **605**, 72 (2005); *Phys. Rev. C* **78**, 035505 (2008).
3. W. Mampe, P. Ageron, C. Bates, J. M. Pendlebury and A. Steyerl, *Phys. Rev. Lett.* **63**, 593 (1989); and update, Ref. [9].
4. W. Mampe, L. N. Bondarenko, V. I. Morozov, Yu. N. Panin and A. I. Fomin, *JETP Lett.* **57**, 82 (1993).
5. J. Byrne *et al.*, *Eur. Phys. Lett.* **33**, 187 (1996).
6. S. S. Arzumanov *et al.*, *Nucl. Instrum. Methods Phys. Res. A* **440**, 511 (2000); and update, *JETP Lett.* **95**, 224 (2012).
7. J. S. Nico *et al.*, *Phys. Rev. C* **71**, 055502 (2005).

8. A. Pichlmaier, V. Varlamov, K. Schreckenbach and P. Geltenbort, *Phys. Lett. B* **693**, 221 (2010).
9. A. Steyerl, J. M. Pendlebury, C. Kaufman, S. S. Malik, A. M. Desai, *Phys. Rev. C* **85**, 065503 (2012).
10. W. Paul *et al.*, *Z. Physik C* **45**, 25 (1989).
11. V. F. Ezhov *et al.*, *J. Res. Mat. Inst. Standards and Technology* **110**, 1 (2005); V. F. Ezhov *et al.*, *Nucl. Instrum. Methods Phys. Res. A* **611**, 167 (2009).
12. K. Leung, O. Zimmer, *Nucl. Instrum. Methods Phys. Res. A* **611**, 181 (2009).
13. P. Huffman *et al.*, *Nature* **403**, 62 (2000).
14. P. L. Walstrom *et al.*, *Nucl. Instrum. Methods Phys. Res. A* **599**, 82 (2009).
15. S. Materne *et al.*, *Nucl. Instrum. Methods Phys. Res. A* **611**, 176 (2009).
16. V. V. Vladimirsky, *JETP* **12**, 740 (1961).
17. Yu. N. Pokotilovski, *JETP Lett.* **76**, 131 (2002); *Erratum, JETP Lett.* **78**, 422 (2003).
18. E. Majorana, *Il Nuovo Cimento* **9**, 43 (1932).
19. J. C. Mallinson, *IEEE Transactions on Magnetics* **9**, 1 (1973).
20. R. W. Pattie *et al.*, *Phys. Rev. Lett.* **102**, 012301 (2009).
21. J. Liu *et al.*, *Phys. Rev. Lett.* **105**, 181803 (2010).
22. A. Steyerl, C. Kaufman, G. Müller, S. S. Malik, and A. M. Desai, *Phys. Rev. C* **86**, 065501 (2012).
23. P. M. Morse and H. Feshbach, *Methods of Theoretical Physics* (McGraw-Hill, New York, 1953), Chap. 9.3.

Vibration-Induced Loss of Ultra-Cold Neutrons in a Magneto-Gravitational Trap

D. J. SALVAT

Department of Physics and Center for Exploration of Energy and Matter, Indiana University, Bloomington, IN 47405, USA

P. L. WALSTROM

Los Alamos National Laboratory, Los Alamos, NM 87545, USA

We investigate the loss of Ultra-Cold Neutrons (UCN) in a magneto-gravitational trap due to microphonic vibrations. The presence of vibrations may cause the spectrum of trapped UCN to shift to an energy higher than the potential energy of the trap, and UCN may therefore escape the trap on a time scale similar to the neutron β decay lifetime of ~ 881 s. For the geometry of the UCNτ trap at Los Alamos National Laboratory (LANL), we present a simplified model of the trap in the presence of vibrations, and observe the magnitude of energy transfers that are imparted on the UCN. From this, we estimate the frequency and amplitude of vibrations that may cause the loss of 10^{-4} of the neutrons over a storage time of 1000 s. We use an accelerometer to measure the frequency and amplitude of microphonic vibrations in the UCN experimental area at LANL, and find them to be similar to those used in the model. Potential implications and future work are discussed.

1. Introduction

The extraction of the neutron lifetime τ from a UCN trap relies on the careful assessment and correction of loss mechanisms aside from neutron β decay. A loss mechanism with a characteristic time of 3 months is enough to shift the observed decay time by $\sigma = 0.1$ s, which is the desired uncertainty of a next generation measurement. With the goal of a 0.1 s measurement error in τ, the consideration of hitherto unexplored loss mechanisms is necessary.

One potential loss channel is that of small energy transfer to a trapped neutron which can subsequently render it energetically untrappable. In this way, the UCN could be slowly "evaporated" away. For both material and magnetic traps, microphonic vibrations of the trap may induce such small energy transfers.

The effect of vibrations on material traps has been discussed in the context of material storage volumes,[1] though not necessarily in the context of precision measurements of the neutron lifetime. In a material storage volume, a UCN can either gain or lose velocity upon bouncing from a material wall depending on the wall's instantaneous velocity $v_{wall} \propto wA$ (where w and A are the angular frequency and amplitude of the vibration). In this way, the UCN can be seen as taking a random walk in velocity with each bounce, so that the net change in velocity is proportional to $\omega A \sqrt{n}$, where n is the number of bounces. However, it is not reasonable to expect that the same behavior would also occur in magnetic traps.

Here, we discuss the effect of microphonic vibrations on the trajectories of UCN in the UCNτ magneto-gravitational trap at the Los Alamos Neutron Science Center (LANSCE).[2] We investigate the trajectory of UCN in a simplified 1-D model that is parametrized by the frequency, amplitude, and phase of oscillation. The size and frequency of microphonic vibrations is then investigated using an accelerometer in the UCN experimental area to assess the potential severity of the effect.

2. Simplified Model

In the simplified model, vibrational motion is 1-D in the vertical (z) direction, the Halbach array is a horizontal plane, and the holding field is neglected. Moreover, ripple terms in the Halbach array field due to the discrete PM blocks are neglected. With no ripple, the magnitude B of the static field in a linear Halbach array of period λ is

$$B(z) = B_0 e^{-kz}, \tag{1}$$

where $k = 2\pi/\lambda$ and B_0 is the field at the surface of the array. When the entire Halbach array is assumed to move up and down vertically in a sinusoidal fashion with amplitude A, angular frequency ω, and phase δ, the field becomes time-dependent and takes the form

$$B(z,t) = B_0 e^{-k[z+A\sin(\omega t+\delta)]} \tag{2}$$

Majorana spin flips are negligible if the inequality

$$\frac{dB/dt}{B} << \frac{\mu|B|}{\hbar} \tag{3}$$

is satisfied everywhere along the orbit. Note that dB/dt has contributions both from the vibration and the motion through field gradients, i.e.,

$$\frac{dB}{dt} = \mathbf{v} \cdot \nabla B + \frac{\partial B}{\partial t} \tag{4}$$

It turns out for reasonable vibrations, the inequality is satisfied. Assuming that the adiabatic criterion is satisfied (i.e., the fractional rate of change of the field along the orbit is much less than the Larmor frequency), the 1-D equations of motion for a field-repelled neutron in this field, with gravity, are

$$\frac{dv_z}{dt} = \frac{k|\mu|}{m} B_0 e^{-k[z+A\sin(\omega t+\delta)]} - g \qquad (5)$$

$$\frac{dz}{dt} = v_z \qquad (6)$$

Since we now have a time-dependent potential, numerical integrators used for tracking in the static potential have to be modified to keep their symplectic properties for long-term tracking. Both modified leapfrog and fourth-order time-dependent symplectic tracking algorithms were used in this work, with satisfactory agreement (see for example the time-dependent prescription in ref. 3 as applied to the optimized fourth-order integrator in ref. 4). An initial set of tracking studies was done to get a sense of the magnitude of vibration effects. Three cases were run with $\lambda = 0.02$ m, $B_0 = 1$ T, $A = 10^{-4}$ m, $f = 100$ Hz ($\omega = 2\pi f$), with phases $0°$, $45°$, and $180°$. The neutron was dropped from $h \equiv z(0) = 0.3$ m with zero velocity, and tracked with a modified leapfrog algorithm for 1 s with 10^8 time steps.

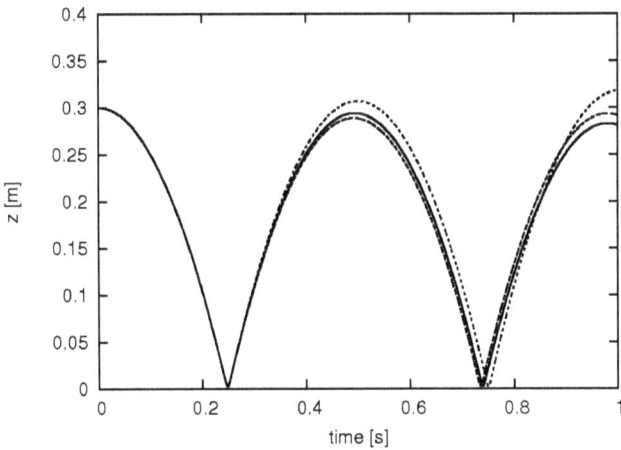

Fig. 1. Plots of z vs. t for various phases. Solid curve: $0°$; long dashes: $45°$, and short dashes: $180°$.

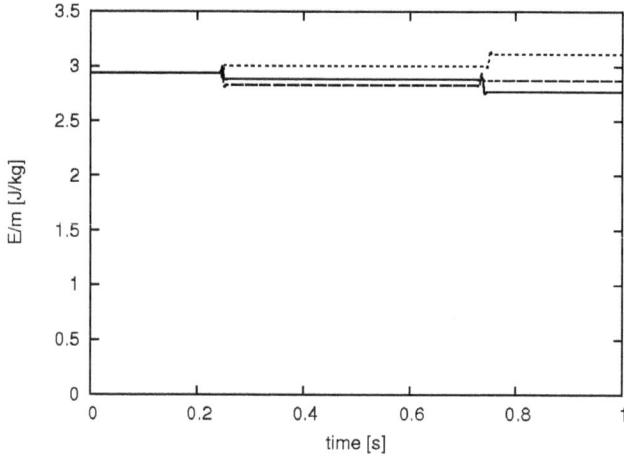

Fig. 2. Plots of E/m *vs.* t for various phases. Solid curve: $0°$; long dashes: $45°$, and short dashes: $180°$.

Plots of z *vs.* t for the three phases are shown in Fig. 1. Note that for phases of $0°$ and $45°$, the energy (i.e. the maximum bounce height) is smaller than the initial energy after 1 s, but for $180°$, the final energy is greater. Figure 2 is a plot of energy/mass for the trajectories of Fig. 1. From this we see that energy transfers of 2% can be observed after one bounce. Though the chosen amplitude of $A = 100$ μm is somewhat generous, it suggests that a more careful sweep of the parameter space would be beneficial to establish the size of typical energy transfers.

3. Parameter Sweep

The parameters h, f, A, and δ are individually varied (with other parameters fixed) to investigate the relationship between the energy transfer per bounce of the UCN. The salient results of the parameter sweep are:

- The energy transfer is roughly linear in E_0.
- The energy transfer obeys $\epsilon \propto \sin(\delta + \phi)$ for some ϕ with all other parameters fixed (see Fig. 3).
- The energy transfer is roughly linear in A (see Fig. 4).
- The energy transfer is maximized for some $f \sim 0 - 100$, and drops off rapidly for higher frequencies (see Fig. 5).

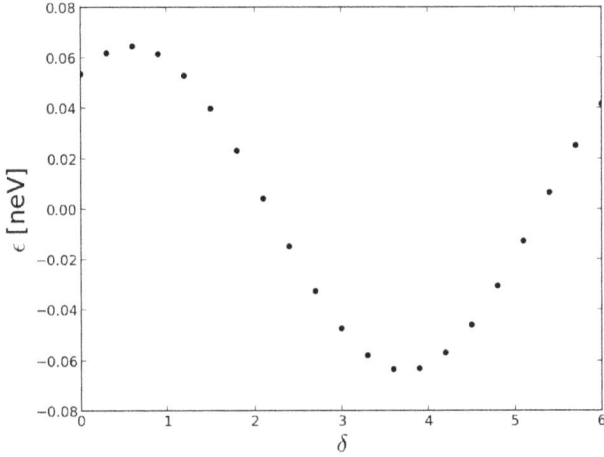

Fig. 3. Energy transfer after 1 bounce as a function of δ to a UCN with $h = 0.44$, $f = 40$, $A = 10^{-5}$.

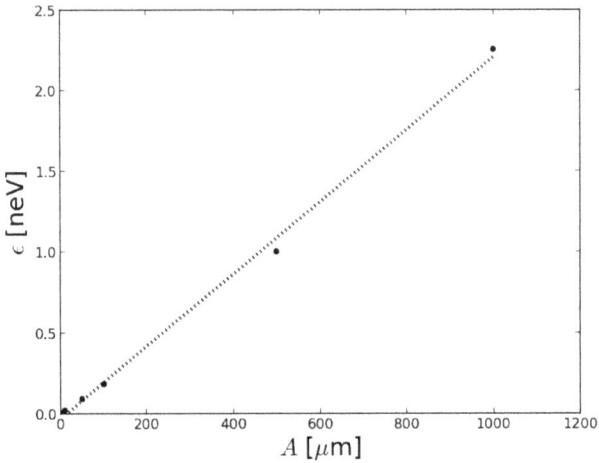

Fig. 4. Energy transfer after 1 bounce as a function of A to a UCN with $h = 0.44$, $f = 40$, $\delta = 2.0$.

This behavior is shown in in figs. 3, 4, and 5. For example, there is a resonant behavior as a function of frequency, perhaps unsurprisingly different from the frequency dependence for material walls. Frequencies above

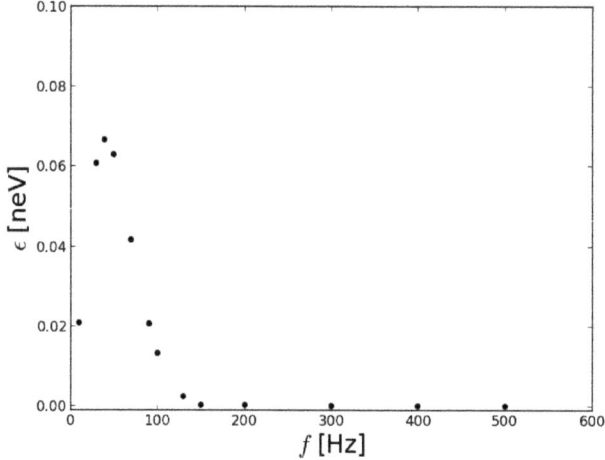

Fig. 5. Energy transfer after 1 bounce as a function of f to a UCN with $h = 0.44$, $\delta = 2.0$, $A = 10^{-5}$.

the few hundred Hz range induce little change in neutron energy, even for large amplitudes.

4. Spectral Random Walk

With the above behavior established, we can estimate the effect of these small UCN energy transfers by randomly walking the initial stored spectrum in a UCN trap, and observing the fraction x of the spectrum that exceeds the maximum trappable energy. In the UCNτ experiment, we intend to remove UCN that reach within approximately 6 cm of the top of the trap, so that the greatest UCN energy is 45.1 neV, while the maximum trappable energy is 50 neV. We estimate a typical number of bounces of a UCN from a previous full 3-D simulation of the trap to be $N = 3000$ over 1000 seconds of storage.

For the random walk computetion, we take an initial spectrum of 3×10^5 UCN, with energy distributed as $\rho(E) \propto \sqrt{E}$ up to $E = 45.1$ neV. Using the observed behavior from section 3, we proceed by shifting each neutron N times as per:

$$E_{i+1} = E_i \pm \epsilon \sin \eta \cdot \frac{E_i}{45.1 \text{ neV}} \tag{7}$$

for each energy. After N iterations, x is computed by counting the number of final energies $E_N > E_t$ where $E_t = 50.3$ neV. We find that $\epsilon = 0.06$ gives $x = 1.5 \cdot 10^{-4}$. The initial and final spectra are shown in Fig. 6.

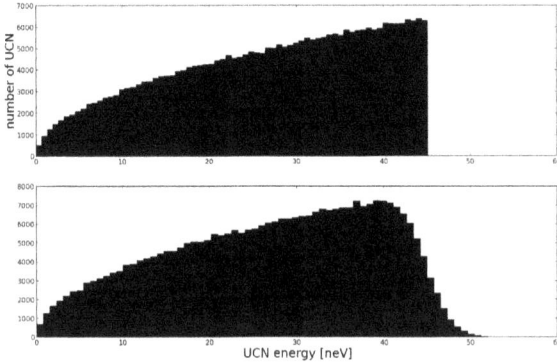

Fig. 6.   The initial (upper) and final (lower) spectra after $N = 3000$ bounces.

From the above, we find the largest value A which satisfies $\epsilon < 0.06$ for a few different frequencies (this is done by hand, so values are approximate). We find for $f = 40$ Hz that $A \approx 8 \cdot 10^{-6}$ m, for $f = 100$ Hz that $A \approx 4 \cdot 10^{-5}$ m, and for $f = 150$ Hz that $A \approx 3 \cdot 10^{-4}$ m. This, then, is a hint of the size of vibrations at these frequencies *over 1000 seconds* that could evaporate more than 10^{-4} of the UCN (note that the loss of 10^{-4} neutrons over 1000 seconds suggests a storage time of roughly 3.8 months not including β decay).

5. Microphonic Vibration Measurements

We can compare the vibrational amplitudes from section 4 to observed vibrations in the UCN experimental area at LANSCE. The presence of, for example, vacuum pumps and compressors may induce large vibrations at low frequencies. To investigate, an STMicroelectronics accelerometer (sensitivity $50\mu g/\sqrt{Hz}$) is fastened firmly to various equipment in the area, including, for example, a vacuum jacket on the UCN beamline used for neutron Electric Dipole Moment (nEDM) research and development.

The device is calibrated to read out acceleration a versus time independently along three axes. From the acquired data $a(t)$ we can readily compute the vibrational Power Spectral Density (PSD) along a particu-

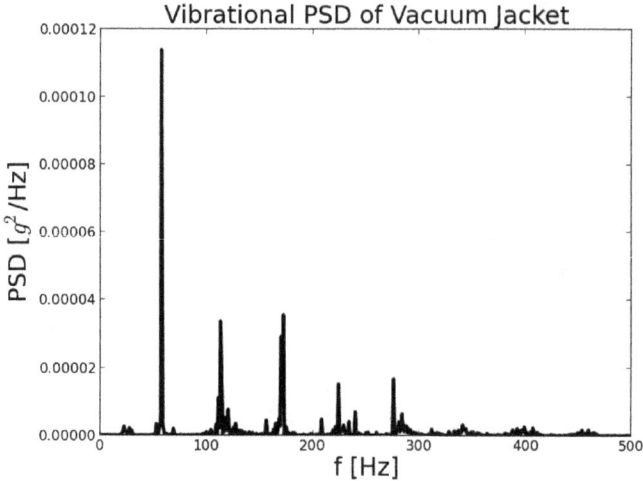

Fig. 7. The PSD along a particular axis of the vacuum jacket on the UCN beamline at LANSCE.

lar direction. Figure 7 shows the PSD along one direction on the nEDM vacuum jacket.

We can estimate the mean amplitude of oscillation at a given frequency via

$$\bar{A} = 2I/\left(2\pi f_0\right)^2, \tag{8}$$

where I is the area under a given peak in the PSD, and f_0 the central frequency of the peak. We identify a few of the most intense low frequency peaks and find $\bar{A} = 16$ μm at 30 Hz to be the largest observed amplitude (measured on the vacuum manifold connecting to the jacket), with more typical values of $\bar{A} = 1$ to 5 μm measured directly on the vacuum jacket.

6. Discussion

While the simple model in section 2 captures some of the qualitative behavior of trapped UCN in the presence of microphonics, UCN trajectories are in general quite complicated, and the presence of quasi-stable orbits (for example) may exacerbate the effect. We also note that the escape time for a UCN with energy $E > E_t$ may not be short compared to the 1000 s storage time, and this is not considered in the above analysis. It therefore may be important to further refine this analysis before considering or ruling out this effect.

The results of section 5 give a sense of the size and frequency content of microphonics in the experimental area. It is, however, unclear how these vibrations translate to that of the actual trap walls, as this depends on geometry and the mechanical coupling of the trap to the rest of the apparatus. This suggests that *in situ* measurements should be done. Problematic sources of vibration could be identified, and given that the effect appears to be more severe at lower frequencies, low frequency mechanical damping can be incorporated into the experimental design. The fraction of UCN which drift to above the maximum trappable energy can also be reduced by lowering the cleaning height. The UCN must drift further in energy in order to evaporate, thereby reducing vibration induced loss. This method comes with the side effect of reducing the effective volume of the trap, and care must be taken to optimize the cleaning procedure.

Finally, it is important to note that the *loss* of neutrons due to vibrational heating is but one potential systematic effect: insofar as UCN detection efficiencies are usually velocity dependent, a change in the UCN spectrum during storage folded with the efficiency of detection at different storage times may induce non-exponential behavior. The subtlety of this issue demands that it be addressed with respect to particular detection methods.

7. Conclusions

A simple model is presented to investigate the trajectories of UCN in the UCNτ trap when subjected to small microphonic vibrations. The model suggests that vibrations of amplitude ~ 10 μm at frequencies below 200 Hz may be able to heat one in ten thousand UCN to an energy above the trap potential over 1000 seconds. Vibrations of similar amplitude and frequency can be observed in the vibrational PSD data acquired from the UCN experimental area at LANSCE. We discuss the potential implications of the result, and conclude that a more realistic analysis, including three-dimensional tracking in the bowl trap with vibrations, is necessary to determine the severity of the effect in the context of a 0.1 s neutron lifetime measurement.

Acknowledgements

Author DJS is supported by the DOE Office of Science Graduate Fellowship Program (DOE SCGF), made possible in part by the American Recovery and Reinvestment Act of 2009, administered by ORISE-ORAU under contract no. DE-AC05-06OR23100.

References

1. R. Golub & J. M. Pendlebury, Rep. Prog. Phys. **42** 439–501 (1979).
2. P. L. Walstrom *et al.*, Nucl. Inst. Meth. in Phys. Res. A **599** No. 1, (2009).
3. R. D. Ruth, IEEE Trans. Nucl. Sci. **NS-30** No. 4 (1983).
4. R.I. McLachlan & P. Atela, Nonlinearity **5** 541–562 (1992).

Blind Analysis in Physics Experiments: Is this Trip Necessary?*

R. GOLUB

Physics Dept, North Carolina State University,
Raleigh, NC26695, USA
rgolub@ncsu.edu

1. Introduction

Blind analysis was introduced into scientific experimentation in order to avoid the problem that in experiments with human subjects subtle clues from the experimenter can influence the response of the subject, initiating the dreaded placebo effect. However even double blinding (where both the subject and the experimenter are unaware of the situation) has proved inadequate where big pharma is involved and it has been necessary to introduce pre-registration of clinical trials as described in.[2] It has now become almost conventional to employ these techniques in physic experiments. According to Richard Feynman[1]

> "It's a thing that scientists are ashamed of- this history- because it's appparent that people did things like this: When they got a number that was too high above Millikan's, they thought something must be wrong - and they would look for and find a reason why something might be wrong. When they got a number closer to Millikan's they didn't look so hard...
> The first principle is that you should not fool yourself - and you are the easiest person to fool."[1]

Peter Galison[3] has given several very interesting examples showing how the decision to end an experiment (i.e. stop searching for systematic errors)

---

*Based on the work of Klein and Roodman[2] we present an alternate conclusion as to the charm of blind analysis in physics experiments.

is influenced by previous known values of the quantity being measured. Blind analysis would evidently help to avoid such situations.

Klein and Roodman[2] have given a detailed argument in support of the need for blind analysis in nuclear and particle experiments and have analyzed several methods of carrying this out. The abstract of their paper summarizes the argument:

"During the past decade, blind analysis has become a widely used tool in nuclear and particle physics experiments. A blind analysis avoids the possibility of experimenters biasing their results toward their own preconceptions by preventing them from knowing the answer until the analysis is completed. There is at least circumstantial evidence that such a bias has affected past measurements, and as experiments become costlier and more difficult and hence harder to reproduce, the possibility of bias has become a more important issue than in the past. We describe here the motivations for performing a blind analysis, and give several modern examples of successful blind analysis strategies."

The goal of this short talk is to discuss some of the arguments in that paper.

2. The Case for Blind Analysis and some Counter-Arguments

Figure 1 is a reproduction of Figure 2 from[2] which shows historical plots of the measurements of 4 quantities. The circles are the results of individual measurements and the dashed lines show the 1σ limits of the published average of the quantity at the time of measurement. The dotted horizontal lines are the values accepted at the time of publication of.[2] The authors calculate that the x^2 value for the hypothesis that the measurements are normally distributed around the previous averages are about 1/2 those associated with the hypothesis that they are normally distributed around the current accepted values. The dramatic shifts in value are seen to coincide with large improvements in accuracy, indicating a radical switch in measurement technique so that experimenter bias is not evident here. The authors themselves conclude:

"Although we cannot say conclusively whether bias has influenced measurements in nuclear and particle physics, the way to avoid even the possibility is to follow Dunnington's and Pfungst's examples and perform measurements while staying blind to the value of our answer."

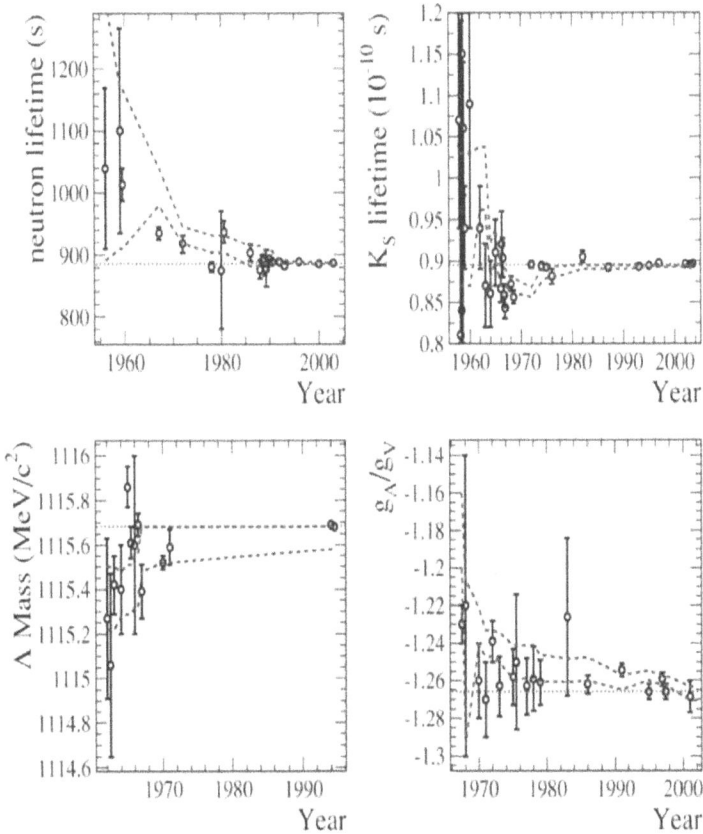

Fig. 1. The history of four measurements compared to published averages before each measurement was made (dashed curves) and the currently accepted value (dotted lines). The space between the dashed curves indicates the 1σ uncertainties on the published values at the time each measurement was made.[2]

However blind analysis has real costs in that it can severely inhibit the search for systematic errors and can preclude studies of unexpected events that occur in 'blinded' regions of the parameter space. To introduce these methods in an attempt to solve a problem that may be non-existent ("..avoid even the possibility ..") seems to the present author to be overkill. The authors give an example of a case where blinding provided a serious obstacle to the performance of an experiment:

"While looking for the decay $\pi^+ \rightarrow e^+\nu$, we focused all our attention on reducing backgrounds, since a prior experiment had set a limit at the level of 10^{-6} on the branching ratio. When we heard that an experiment at CERN had seen a signal around 10^{-4}. I switched from delayed to prompt. The signal was right there, and could have been seen on the first day (B. Richter, private communication)."

In addition

"We note that none of the blind techniques we describe here—and perhaps no blind technique—can be applied to an analysis in which backgrounds are cut or signals identified by event-by-event human inspection."

Again quoting from[2]

"...what to do if the analysis breaks down...It is not necessary in the blind analysis approach to insist that, because an analysis was done blindly, no additional selections may be applied....The blind analysis method does not require that data analysis stop after unblinding, nor does it ensure that the results of the analysis are correct. There is no reason to publish an analysis known to be wrong just because the analysis was done blindly.
Multiple independent analyses are occasionally suggested as a way to prevent experimenter's bias."

In describing a use of blinding in a measurement of the gravitational force:

"The Irvine group's measurement relied on precise knowledge of many different detector parameters—the dimensions of the torsion balance and test masses, the positions of the test masses, and of course the masses of all test components. To prevent themselves from selecting data in a biased way, or from (in their words) "slackening of analysis effort" when their answer began to meet their expectations (what we have called a stopping bias), they kept the value of their near mass known only to 1%—the exact mass known only to someone outside their collaboration. They used the true value of the mass only when they had completed the analysis and were ready to report their initial results. Subsequent improvements to the analysis were made and later published, but they nevertheless published the measurement made before these improvements were made."

So that it appears that at least in this case "an analysis known to be wrong" was published "just because the
analysis was done blindly".

Further

"The next quandary may occur if there are more events in the signal box than expected from backgrounds, but the events are *very inconsistent with the expected signal properties.*" (emphasis added)

So it appears preconceived notions cannot be completely banished.

Klein and Roodman give another example of a case of blind analysis:

"KTeV used a hidden offset directly in its ε'/ε fit. Instead of fitting for the value of ε'/ε, the fit used

$$\varepsilon'/\varepsilon(Hidden) = \left\{ \begin{matrix} + \\ - \end{matrix} \right\} 1 \times \varepsilon'/\varepsilon + C \tag{1}$$

where C was a hidden random constant, and the choice of 1 or -1 was also hidden and random.
The +1 or −1 in the hidden value served to hide the direction ε'/ε changed as different corrections or selections were applied (48). In practice, KTeV had to remove the sign choice at an earlier stage to permit a full evaluation of systematic errors. Nevertheless, the first KTeV ε'/ε result was unblinded only one week before the result was made public.
The addition of an unknown sign also hides the direction the result has moved with changes to the analysis."

So it appears that overzealous blinding (of the sign) was incompatible with a proper consideration of systematic errors.

3. Discussion

Based on the paper by Klein and Roodman[2] written in support of the use of blind analysis in physics experiments we have attempted to show that the method, introduced into science by medical researchers to solve a real problem, has little place in physics research, since, as the authors admit there is surprisingly little evidence for experimenter bias in physics research even taking into account the cases cited by Feynman[1] and Galison[3] and the introduction of the technique has real costs in both the ability to study systematic errors that are often unknown at the planning stage of the experiment and come to light during the measurement and analysis and the ability to follow up unexpected results which may only show up after opening the box.

The experience in medical research has shown that even double blind analysis is not sufficient to avoid experimental bias when the experimenters are really determined. Physicists on the other hand still have the ability to follow the advice of my mentor, Prof Gerrold R. Zacharias, who often repeated the statement that experimental physics was really all about Character.

References

1. Feynman, R., 'Surely You're joking Mr. Feynman' , Norton (1985) as quoted in (2).
2. Klein, J. and Roodman, A., 'Blind analysis in nuclear and particle physics' Ann. Rev. Nucl. Part. Sci. **55**, 141-63, (2005).
3. Gallison, Peter, 'How experiments end', Univ. Chicago Press (1987).

A Technique For Determining Neutron Beam Fluence to 0.01 % Uncertainty

A. T. YUE

Institute for Research in Electronics and Applied Physics, University of Maryland,
College Park, MD 20742, USA
and
National Institute of Standards and Technology
Gaithersburg, MD, 20899, USA
ayue@nist.gov

M. S. DEWEY, D. M. GILLIAM, and J. S. NICO

National Institute of Standards and Technology
Gaithersburg, MD, 20899, USA

N. FOMIN

Los Alamos National Laboratory
Los Alamos, NM, 87545, USA

G. L. GREENE

Physics Department, University of Tennessee,
Knoxville, TN 37996, USA
and
Oak Ridge National Laboratory
Oak Ridge, TN, 37831, USA

W. M. SNOW

Physics Department, Indiana University,
and
Center for Exploration of Energy and Matter
Bloomington, IN 47408, USA

F. E. WIETFELDT

Physics Department, Tulane University,
New Orleans, LA 70118, USA

The achievable uncertainty in neutron lifetime measurements using the beam technique has been limited by the uncertainty in the determination of the neutron density in the decay volume. In the Sussex-ILL-NIST series of beam lifetime experiments, the density was determined with a neutron fluence monitor that detected the charged particle products from neutron absorption in a thin layer of ^{6}Li or ^{10}B. In each of the experiments, the absolute detection efficiency of the neutron monitor was determined from the measured density of the neutron absorber, the thermal neutron cross section for the absorbing material, and the solid angle of the charged particle detectors. The efficiency of the neutron monitor used in the most recent beam lifetime experiment has since been measured directly by operating it on a monochromatic neutron beam in which the total neutron rate is determined with a totally absorbing neutron detector. The absolute nature of this technique does not rely on any knowledge of neutron absorption cross sections or a measurement of the density of the neutron absorbing deposit. This technique has been used to measure the neutron monitor efficiency to 0.06 % uncertainty. We show that a new monitor and absolute neutron detector employing the same technique would be capable of achieving determining neutron fluence to an uncertainty of 0.01 %.

Keywords: Neutron lifetime; neutron fluence

1. Introduction

In the Sussex-ILL-NIST beam lifetime experiments,[1–5] the absolute rate of neutron decays in a volume of the beam was measured by trapping and counting decay protons and the neutron fluence was determined with a neutron monitor that counted the charged particle products produced by the absorption of the beam in a thin deposit of ^{6}Li or ^{10}B. In the most precise of these measurements,[4,5] the neutron fluence was determined by counting the 2070 keV alphas and 2720 keV tritons emitted from neutron absorption in a layer of ^{6}LiF evaporated on a single-crystal silicon wafer. A rigid frame held the thin ($\bar{\rho} = 39.3\ \mu\mathrm{g/cm}^2$) ^{6}LiF target deposit fixed with respect to four precision ground apertures. The apertures masked four charged particle detectors, defining the solid angle for detection of the reaction alphas and tritons. The neutron monitor is characterized by an efficiency parameter ϵ that denotes the ratio of detected reaction products to incident neutrons. The efficiency of the monitor is dependent on the velocity and intensity distribution of the beam it is operated on, so a reference efficiency is defined from which the efficiency for any beam may be in princple determined. The reference efficiency of the monitor $\epsilon_0(0,0)$ is defined for an infinitely narrow (i.e. intensity distribution $\phi(x,y) = \delta(x)\delta(y)$) monochromatic beam of thermal neutrons ($v = v_0 = 2200$ m/s, corsponding to $\lambda_0 = 0.1798$ nm) incident on the center of an infinitely thin deposit of areal density $\rho(0,0)$:

$$\epsilon_0(0,0) = \frac{2N_A}{A}\sigma_0\Omega_{\text{FM}}(0,0)\rho(0,0), \tag{1}$$

where N_A is the Avogadro constant, $A = 6.01512$ g/mol is the atomic weight of ^{6}Li, σ_0 is the ^{6}Li(n,t)^{4}He thermal neutron cross section, and $\Omega_{\text{FM}}(0,0)$ is the monitor solid angle (as a fraction of 4π) for a particle emitted from the center of the deposit.

In the NIST experiment, the reference efficiency was originally obtained from the measured solid angle (0.1 % uncertainty, where "uncertainty" here and throughout the document corresponds to the standard 1 σ uncertainty), the measured deposit areal density (0.25 % uncertainty), and the ENDF/B-VI ^{6}Li(n,t)^{4}He thermal neutron cross section (0.14 % uncertainty).[6] The 0.3 % uncertainty attained with this method is the dominant source of uncertainty in the experiment and is likely near the precision limits of the techniques used. Additionally, the efficiency is dependent on the results of a global analysis of a large collection of neutron cross section experiments. Since the publication of the most recent NIST lifetime result, a new evaluation of the ^{6}Li(n,t)^{4}He thermal neutron cross section has been published,[7] lowering the cross section (and thus τ_n) by 0.25 %.

Considerable time and effort have been spent developing methods to directly measure the detection efficiency of the monitor.[8–15] Direct measurement of the neutron monitor efficiency is performed by measuring the rate of alphas and tritons on a monochromatic neutron beam of well-measured wavelength λ_{mono} and using a second, totally absorbing ("black") neutron detector to determine the total rate of neutrons passing through the monitor. In such an experiment, the observed rate of alphas and tritons ($r_{\alpha,t}$) is

$$r_{\alpha,t} = \epsilon R_n = \epsilon_0 \frac{\lambda_{\text{mono}}}{\lambda_0} R_n, \tag{2}$$

where ϵ is the detection efficiency of the monitor for neutrons of wavelength λ_{mono} and ϵ_0 is the efficiency of the monitor for an equivalent beam of thermal neutrons. The experimental challenge is to measure both λ_{mono} and R_n. This was recently accomplished with an alpha-gamma counter (known as the Alpha-Gamma device) that used high-purity germanium (HPGe) detectors to count the 478 keV reaction gamma rays from neutron absorption in a totally absorbing ^{10}B target.[9,10,15] The HPGe detectors were calibrated through a series of transfer calibrations from an alpha source of known absolute activity. The neutron monitor efficiency was measured to 0.06 % precision.[15] The measurement was statstics-limited and the largest sources of systematic uncertainty have the potential to be significantly reduced. In this paper, we investigate the possibility of constructing a new black detector based on the Alpha-Gamma technique and a new neutron monitor to measure neutron fluence to 0.01 % uncertainty for the purposes of a 0.01 % neutron lifetime measurement.

2. The Alpha-Gamma Technique

The Alpha-Gamma technique uses HPGe detectors and passivated implanted planar silicon (PIPS) detectors to count alpha and gamma radiation emitted from radioactive sources and neutron absorbing targets placed in an interchangeable target holder. An alpha-to-gamma cross calibration procedure is used to determine the number of neutrons absorbed in the totally-absorbing ^{10}B target per detected 478 keV gamma ray.

The HPGe detector calibration procedure begins with the determination of the absolute activity of an alpha source in a low-solid angle counting stack.[16,17] The counting stack consists of a spacer cylinder and a diamond-turned aperture. The aperture diameter is measured with a coordinate measuring machine and the distance between the source and the plane of the defining edge of the aperture is measured with a coordinate measuring microscope. From these dimensions, the solid angle of the counting stack is calculated. The absolute activity of the source R_{Pu} is determined from the measured alpha rate $r_{\mathrm{Pu}}(\mathrm{stack})$ and the solid angle Ω_{stack}

$$R_{\mathrm{Pu}} = \frac{r_{\mathrm{Pu}}(\mathrm{stack})}{\Omega_{\mathrm{stack}}}. \tag{3}$$

The source is then placed in the Alpha-Gamma device target position. The measured count rate $r_{\mathrm{Pu}}(\mathrm{AG})$ and the known absolute activity R_{Pu} determine the solid angle of the Alpha-Gamma PIPS detectors Ω_{AG}

$$\Omega_{\mathrm{AG}} = \frac{r_{\mathrm{Pu}}(\mathrm{AG})}{R_{\mathrm{Pu}}}. \tag{4}$$

The source is replaced by a thin ($\rho \approx 20~\mu\mathrm{g/cm}^2$) deposit of enriched ^{10}B prepared in the same fashion as the ^{6}LiF deposit. A monochromatic neutron beam of total neutron rate R_n is turned on and allowed to strike the deposit. The rate of neutron absorption in the deposit $r_n(\mathrm{thin})$ is given by

$$r_n(\mathrm{thin}) = R_n \sigma \rho \frac{N_A}{A}, \tag{5}$$

where σ is the ^{10}B absorption cross section, ρ is the areal density of ^{10}B in the deposit, and $A = 10.12937$ g/mol is the atomic weight of ^{10}B. Neutron absorption in ^{10}B produces a ^{7}Li nucleus and an alpha. The ^{7}Li nucleus is in an excited state 93.70 % of the time[18,19] and rapidly de-excites ($\tau = 73$ fs) by emission of a gamma ray. The alpha-only reaction produces a 1776 keV alpha and the alpha + gamma reaction produces a 1472 keV alpha particle and a 478 keV gamma ray. The emitted alpha particles and gamma rays are detected in the PIPS and HPGe detectors respectively. The observed rate of alpha particles $r_\alpha(\text{thin})$ is

$$r_\alpha(\text{thin}) = \Omega_{\text{AG}} r_n(\text{thin}), \tag{6}$$

and the observed rate of 478 keV gamma rays $r_\gamma(\text{thin})$ is

$$r_\gamma(\text{thin}) = \epsilon_\gamma b_{\alpha\gamma} r_n(\text{thin}), \tag{7}$$

where ϵ_γ is the HPGe detection efficiency for the 478 keV gamma rays emitted from the target and $b_{\alpha\gamma} = 0.9370$ is the branching ratio for the alpha + gamma reaction. The observed alpha rate and the measured solid angle of the PIPS detectors are used to determine the neutron absorption rate

$$r_n(\text{thin}) = \frac{r_\alpha(\text{thin})}{\Omega_{\text{AG}}}. \tag{8}$$

This is, in turn, used to determine ϵ_γ

$$\epsilon_\gamma = \frac{r_\gamma(\text{thin})}{b_{\alpha\gamma} r_n(\text{thin})} = \frac{1}{b_{\alpha\gamma}} \frac{r_\gamma(\text{thin})}{r_\alpha(\text{thin})} \Omega_{\text{AG}}. \tag{9}$$

The thin ^{10}B deposit is replaced with a thick ^{10}B$_4$C target enriched to 98 % ^{10}B. To excellent approximation (>99.99 %), the entire beam is absorbed by ^{10}B and the observed gamma rate is

$$r_\gamma(\text{thick}) = \epsilon_\gamma b_{\alpha\gamma} R_n. \tag{10}$$

Using Eq. 9, R_n is expressed entirely in terms of observables, without any reference to $b_{\alpha\gamma}$

$$R_n = \frac{r_\gamma(\text{thick})}{\epsilon_\gamma b_{\alpha\gamma}} = r_\gamma(\text{thick}) \frac{r_\alpha(\text{thin})}{r_\gamma(\text{thin})} \frac{1}{\Omega_{\text{AG}}}. \tag{11}$$

The thermal neutron detection efficiency of the monitor ϵ_0 is then

$$\epsilon_0 = \frac{r_{\alpha,t}}{r_\gamma(\text{thick})} \frac{r_\gamma(\text{thin})}{r_\alpha(\text{thin})} \Omega_{\text{AG}} \frac{\lambda_0}{\lambda_{\text{mono}}}. \tag{12}$$

Ancillary measurements of systematic effects are performed to obtain the reference efficiency $\epsilon_0(0,0)$.

3. Uncertainty Achieved With the Existing Devices

The Alpha-Gamma technique has been used to perform a 0.06 % measurement of $\epsilon_0(0,0)$.[15] The summary of uncertainties is shown in Table 1. In order to perform a 0.01 % measurement with this technique, six sources of uncertainty must be reduced.

Table 1. Sources of uncertainties in the Alpha-Gamma measurement of $\epsilon_0(0,0)$ and their fractional uncertainties.[15]

Source of uncertainty	Fractional uncertainty
Neutron counting statistics	3.1×10^{-4}
α-source calibration of AG α-detector	2.7×10^{-4}
γ attenuation in B_4C target	2.5×10^{-4}
Neutron beam wavelength	2.4×10^{-4}
γ attenuation in thin ^{10}B target	1.3×10^{-4}
$\frac{\lambda_{mono}}{2}$ contamination of beam	1.0×10^{-4}
Neutron backscatter in monitor substrate	3.9×10^{-5}
AG α solid angle for beam spot	2.7×10^{-5}
Detector dead time	2.4×10^{-5}
Neutron loss in Si substrate	1.8×10^{-5}
Neutron absorption by ^6Li	1.2×10^{-5}
Self-shielding of ^6Li deposit	6.0×10^{-6}
Neutron monitor solid angle for beam spot	4.5×10^{-6}
γ production in thin ^{10}B target Si subtrate	3.2×10^{-6}
Monitor misalignment w.r.t. beam	2.0×10^{-6}
Neutron scattering from B_4C	3.3×10^{-7}
Total	5.7×10^{-4}

The Alpha-Gamma experiment was limited by statistical uncertainty attributable to charged particle counting in the neutron monitor (0.031 %) and alpha particle counting with the ^{239}Pu source (0.027 %). The neutron monitor rate ($r_{\alpha,t} \approx 15$ s^{-1}) was limited by the low detection efficiency of the monitor ($\epsilon \approx 9 \times 10^{-5}$) and strong restrictions on the beam diameter due to the limited acceptance of the Alpha-Gamma target (accepts a maximum beam diameter of 22 mm compared to 38 mm for the monitor). The ^{239}Pu source measurements were limited by the observed alpha rates, attributable to the total source activity ($R_{Pu} \approx 24000$ s^{-1}) and the low solid angle of the detection geometries ($\Omega_{stack} \approx 5 \times 10^{-3}$ and $\Omega_{AG} \approx 7 \times 10^{-3}$).

The neutron beam wavelength λ_{mono} (and a small $\frac{\lambda_{mono}}{2}$ component not removed by an upstream beryllium filter) were measured by Bragg scattering with a perfect crystal silicon analyzer. Four of the thirty-one measurements of the Bragg angle fell outside the expected normal distribution of measurements about the mean.[15] As such, the 1-σ measurement uncertainty was conservatively estimated to be the difference between the extremum and the mean of the data set (0.024 %). The correction for the presence of a small $\frac{\lambda_{mono}}{2}$ component of the beam was determined in one measurement with an uncertainty of 0.01 %.

The orientation of the HPGe detectors in the Alpha-Gamma device is such that gamma rays produced by neutron absorption in the thin and thick ^{10}B targets must travel through the target in order to reach one of the two HPGe detectors. Therefore, attenuation corrections are required for accurate determination of r_γ(thin) and r_γ(thick). The attenuation in each target was determined by stacking additional targets and measuring the gamma rate. The uncertainty in these corrections comes from statistical uncertainty in the gamma rate measurement and, for the r_γ(thick) correction, the uncertainty in the measurement of the thickness of the ^{10}B$_4$C targets.

4. New Devices for a Measurement of Neutron Fluence to 0.01 % Uncertainty

The statistical uncertainty can be reduced by constructing new devices capable of accepting a much larger beam. The candidate devices (referred to as the updated neutron monitor and updated Alpha-Gamma device) shown in Fig. 1 use 60 mm ^{6}LiF and ^{10}B deposits (up from 38 mm) on 76 mm backings (up from 50 mm), which allows for the size of the beam at the downstream collimator to be increased from 10.5 mm to 35 mm. Each device measures charged particle reaction products with four PIPS detectors masked by apertures such that $\Omega_{FM} = \Omega_{AG} \approx 4.2 \times 10^{-3}$. Through careful choice of the angle the apertures face the target deposit, the detection solid angle of each device can be made maximally insensitive to shape of the beam profile. To achieve reasonable detection rates in the updated neutron monitor on a high-fluence polychromatic beam (as used in the beam lifetime experiment) and a monochromatic beam (as used in the Alpha-Gamma experiment), the areal density of the ^{6}LiF target deposit is set to 5 μg/cm^2 (is 39.3 μg/cm^2 in existing device). The areal density of the thin ^{10}B deposit in the updated Alpha-Gamma device is set to 40 μg/cm^2 (is $\approx$ 20 μg/cm^2 in the existing device). On a monochromatic ($\lambda = 0.496$ nm) beam with a neutron fluence of 1×10^6 cm^{-2}s^{-1}, we anticipate $r_{\alpha,t} \approx 100$ s^{-1} and r_α(thin) ≈ 1000 s^{-1}. In the updated Alpha-Gamma device, four front-facing 10 % relative efficiency HPGe detectors will be arranged in identical fashion to the PIPS detectors such that $\epsilon_\gamma \approx 5 \times 10^{-4}$. The arrangement of the HPGe detectors eliminates the large corrections for gamma attenuation in the targets. The expected gamma rates from the thin and thick targets are $r_\gamma(\text{thin}) \approx 100 \text{ s}^{-1}$ and $r_\gamma(\text{thick}) \approx 4000 \text{ s}^{-1}$.

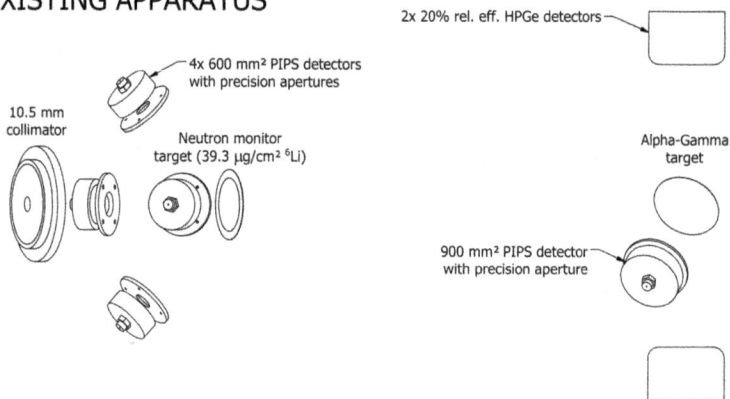

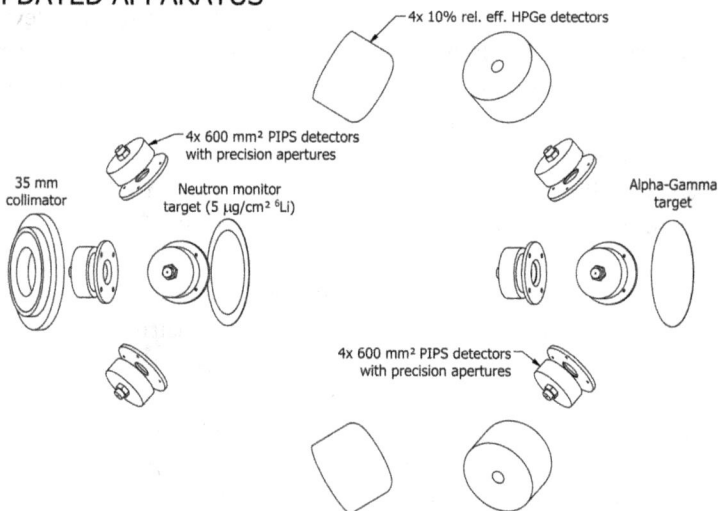

Fig. 1. Drawings of the existing neutron monitor and Alpha-Gamma device and possible updated versions of the devices.

To determine the run time required to reach $0.01\,\%$ statistical uncertainty, we assume the same run schedule used in the existing Alpha-Gamma experiment.[15] Each measurement of the efficiency of the neutron monitor would consist of a one-day measurement of $\frac{r_{\alpha,t}}{r_\gamma(\text{thick})}$ bracketed by two one-day measurements of $\frac{r_\alpha(\text{thin})}{r_\gamma(\text{thin})}$ to eliminate possible linear drift in ϵ_γ. At the expected signal rates, it is estimated that 36 3-day efficiency measurements would be necessary to achieve 0.01 % statistical uncertainty.

To determine the total time required to perform the Pu source calibration, we assume $\Omega_{\text{stack}} \approx 5 \times 10^{-3}$ as in the existing source calibration device but use a source with absolute activity $R_{\text{Pu}} \approx 10^5$ s^{-1}. The anticipated rate is $r_{\text{Pu}}(\text{stack}) \approx 500$ s^{-1}, and a statistical uncertainty of 0.002 % can be achieved in 60 days of accumulation (which could be performed in parallel to the neutron measurements). The solid angle for the Pu source calibration stack has been measured to 0.007 % uncertainty,[17] thus R_{Pu} can be determined to 0.0073 % uncertainty in this period of time. The expected rate in the Alpha-Gamma alpha detector is $r_{\text{Pu}}(\text{AG}) \approx 420$ s^{-1}, thus Ω_{AG} could then be determined to 0.008 % uncertainty in approximately 30 days of counting. This accumulation could be performed during routine maintenance of the neutron source (e.g., reactor downtime) to minimize the total time required to complete the measurements.

The wavelength measurement technique has been demonstrated previously to be capable of determining λ_{mono} to a precision of 0.006 % and the correction for the $\frac{\lambda_{\text{mono}}}{2}$ component to 0.001 %.[13] Systematic uncertainties due to alignment of the wavelength measuring device can be controlled to the 0.001 % level.[20] Additional measurement time would be required with the existing device to determine the cause of the discrepant data points, and additional measurements of the $\frac{\lambda_{\text{mono}}}{2}$ component would be required to reduce the measurement uncertainty to the level demonstrated previously.

With these improvements, the new experiment would be capable of measuring the updated neutron monitor efficiency to 0.014 % in a comparable amount of time to that used to perform the 0.06 % measurement of the existing neutron monitor efficiency.

5. Conclusions

The Alpha-Gamma technique has been used to measure the absolute efficiency of a neutron monitor based on neutron absorption in ^{6}Li to 0.06 % uncertainty. A review of the uncertainty budget shows that six sources of uncertainty must be reduced in order to measure the detection efficiency of a neutron monitor to 0.01 % uncertainty. We show that these uncertainties can be addressed by construction of an updated neutron monitor and

Alpha-Gamma device and with a longer and more precise set of beam wavelength measurements. We find that the new neutron fluence measurement could be performed without changes to the existing technique and the devices could, in principle, be constructed without additional research and development.

References

1. J. Byrne, J. Morse, K. F. Smith, F. Shaikh, K. Green and G. L. Greene, *Physics Letters B* **92**, 274 (1980).
2. J. Byrne, P. G. Dawber, J. A. Spain, A. P. Williams, M. S. Dewey, D. M. Gilliam, G. L. Greene, G. P. Lamaze, J. Pauwels, R. Eykens and A. Lamberty, *Physical Review Letters* **65**, 289 (1990).
3. J. Byrne, P. G. Dawber, C. G. Habeck, S. J. Smidt, J. A. Spain and A. P. Williams, *Europhys. Lett.* **33**, 187 (1996).
4. M. S. Dewey, D. M. Gilliam, J. S. Nico, F. E. Wietfeldt, X. Fei, W. M. Snow, G. L. Greene, J. Pauwels, R. Eykens, A. Lamberty and J. VanGestel, *Physical Review Letters* **91**, p. 152302 (2003).
5. J. S. Nico, M. S. Dewey, D. M. Gilliam, F. E. Wietfeldt, X. Fei, W. M. Snow, G. L. Greene, J. Pauwels, R. Eykens, A. Lamberty, J. VanGestel and R. D. Scott, *Physical Review C* **71**, p. 055502 (2005).
6. A. D. Carlson, W. P. Poentiz, G. M. Hale, R. W. Peelle, D. C. Doddler, C. Y. Fu and W. Mannhart, *NIST Interagency Report* **5177** (1993).
7. A. D. Carlson *et al.*, *Nucl. Data Sheets* **110**, 3215 (2009).
8. G. P. Lamaze, D. M. Gilliam and A. P. Williams, *Journal of Radioanalytical and Nuclear Chemistry, Articles* **123**, 551 (1988).
9. D. M. Gilliam, G. L. Greene and G. P. Lamaze, *Nuclear Instruments & Methods in Physics Research, Section A* **284**, 220 (1989).
10. J. M. Richardson, Accurate Determination of Thermal Neutron Flux: A Critical Step in the Measurement of the Neutron Lifetime, PhD thesis, Harvard University (1993).
11. Z. Chowdhuri, Absolute Neutron Measurements in Neutron Decay, PhD thesis, Indiana University (2000).
12. Z. Chowdhuri, G. L. Hansen, V. Jane, C. D. Keith, W. M. Lozowski, W. M. Snow, M. S. Dewey, D. M. Gilliam, G. L. Greene, J. S. Nico and A. K. Thompson, *Review of Scientific Instruments* **74**, 4280 (2003).
13. G. L. Hansen, A Radiometric Measurement of Neutron Flux in a Liquid ^{3}He Target, PhD thesis, Indiana University (2004).
14. F. E. Wietfeldt, Absolute Thermal Neutron Fluence Measurement Using ^{3}He Gas Scintillation (unpublished, 2009).
15. A. T. Yue, Progress Towards a Redetermination of the Neutron Lifetime Through the Absolute Determination of Neutron Flux, PhD thesis, University of Tennessee (2011).
16. D. M. Gilliam, B. Denecke, R. Eykens, J. Pauwels, P. Robouch, P. Hodge, J. M. R. Hutchinson and J. S. Nico, *Nuclear Instruments & Methods in Physics Research, Section A* **438**, 124 (1999).

17. D. M. Gilliam and A. T. Yue, Improvements in the Characterization of Actinide Targets by Low Solid-Angle Counting, *Journal of Radioanalytical and Nuclear Chemistry*. Available online: doi:10.1007/s10967-013-2647-z
18. A. J. Deruytter and P. Pelfer, *Journal of Nuclear Energy* **21**, 833 (1967).
19. M. L. Stelts, R. E. Chrien, M. Goldhaber and M. J. Kenny, *Physical Review C* **19**, 1159 (1979).
20. K. J. Coakley, M. S. Dewey, A. T. Yue and A. B. Laptev, *Nuclear Instruments & Methods in Physics Research, Section A* **611**, 293 (2009).

A New Method of Neutron Detecton for UCN Lifetime Measurements[*]

C. L. MORRIS[†], D. J. SALVAT[‡], E. ADAMEK[‡], D. BOWMAN[§], S. CLAYTON[†],
C. CUDE[‡], W. FOX[‡], G. HOGAN[†], K. HICKERSON[**], A. T. HOLLEY[‡], C.-Y. LIU[‡],
M. MAKELA[†], G. MANUS[‡], S. PENTTILA[§], J. RAMSEY[†], A. SAUNDERS[†],
S. SAWTELLE[‡], S. J. SEESTROM[†], K. SOLBERG[‡], J. VANDERWERP[‡],
B. VORNDICK[††], P. WALSTROM[†], Z. WANG[†], A. R. YOUNG[††]

A number of inconsistent neutron lifetime measurements have been reported. The disagreement among the various measurements made with material neutron traps with ultra-cold neutrons (UCN) suggests unaccounted for systematic errors in these measurements. One potential source of error is due to the long emptying times which may be time dependent due to the UCN phase space evolution in the trap. We present a way to reduce this effect.

1. Introduction

Recent values reported for the neutron lifetime, Fig. 1, reveal more than 5 standard deviations of disagreement between the most precise reported values. This situation leads the Particle Data Group (PDG) to express their frustration with: "There seems little better to do than to again average the best seven measurements. The result, 880.1 ± 1.1 s (including a scale factor of 1.8), is 5.6 s lower than the value we gave in 2010—a drop of 7.0 old and 5.1 new standard deviations."[1]

Most previous bottle measurements have counted the surviving neutrons after storage by draining them into an external detector. Typical drain times have been much larger than the stated total uncertainties. It is possible that evolution of neutron phase space in the trap can change the draining time and affect the neutron detection efficiency as a function of storage time and thus

[*] This work is supported by U.S. Department of Energy Office of Science
[†] Los Alamos National Lab
[‡] Indiana University
[§] Oak Ridge National Lab
[**] California Institute of Technology
[††] North Carolina State University

provide an unanticipated source of systematic error. Although the motivation for further measurements is clear, there is also a need for new experiemental methods that can reduce the sources of systematic error that have resulted in this difficult situation.

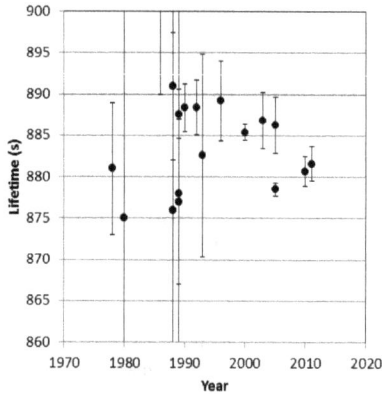

Fig. 1. Recent values reported for the neutron lifetime taken from Ref. [1].

2. Los Alamos UCN Lifetime Experiment

We have begun work on a new lifetime experiment in Los Alamos using the Los Alamos UCN source,[2] depicted in Fig. 2, to load a large permanent magnet trap.[3] The source provides a density of 50 unpolarized UCN/cc and 20 polarized UCN/cc with energies below 180 neV.

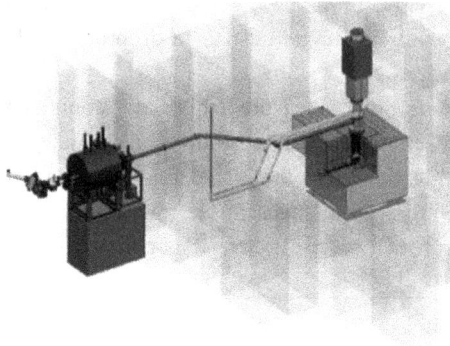

Fig. 2. Los Alamos UCN Source.

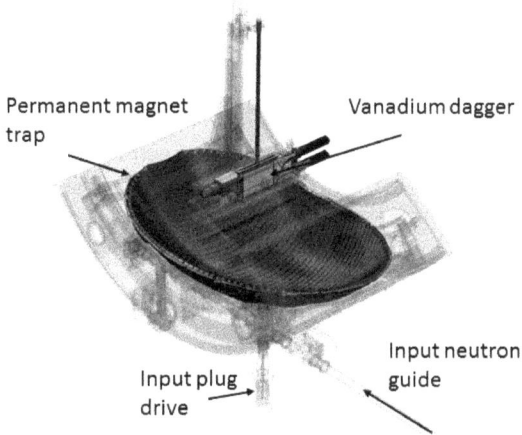

Fig. 3. Schematic of the Los Alamos UCNtau experiment with some of the major features labeled. Not shown are holding field coils that provide a magnetic field orthogonal to the trapping field from the permanent magnet Halbach array.

The trap has been designed to provide a large volume, a short cleaning time, and a long depolarization time. The trap is loaded through a section of the permanent magnet wall that can be lowered to provide a path into the bottom of the trap.

In this report we present a new method for counting ultra-cold neutrons (UCN) in place within the trap that may reduce the systematic errors typically associated with the very long unloading times that are associated with UCN bottles.

2.1. *Neutron Capture on Vanadium*

Nearly all materials in the periodic table have positive, repulsive, Fermi potential. The Fermi potential is given by $U_F = \dfrac{2\pi\hbar^2}{m} Nb$, where $\hbar$ is the reduced Plank constant, m is the neutron mass, N is the number density of the material, and b is the neutron scattering length.

Of the exceptions, vanadium has unique properties that make it ideal for counting UCN.[4] Firstly, it has a small negative potential of $U_F \approx -7$ neV, so there is very little reflection from its surface. Secondly, natural vanadium is nearly mono-isotopic ^{51}V with a neutron capture cross section of 5.08 b. The lifetime of neutrons in the material is $\tau = 13$ μs, so a fraction of a mm absorbs all neutrons that enter. Finally, neutron capture leads to ^{52}V, which β decays with a relatively high endpoint energy of 2.54 MeV accompanied nearly 100 %

of the time by a 1.43 MeV γ-ray, with a lifetime of 5.40 minutes. The β and γ rays can be detected in coincidence to reduce backgrounds.

One method for measuring the neutron lifetime is to load a trap, clean the trap to remove super barrier neutrons, store, and then count the remaining neutrons after suitable storing times. Another method is to count the neutron decay products and measure the decay rate vs. time. Although the latter method is statistically more efficient, we have chosen the former method for our initial experiments because: 1) detecting decay electrons suffers from large, potentially time-dependent, backgrounds from Compton electrons produced by γ-rays interacting in the detectors and surrounding materials; and 2) we have not found a way to detect either electrons or protons that doesn't have lower efficiency for decays that occur near the trap walls and we therefore worry about time dependence introduced by phase space evolution in such detection schemes.

2.2. *Los Alamos Lifetime Experiment*

Our proposed method of neutron counting is illustrated in Fig. 4. A counting sequence starts by lowering a vanadium foil into the trap. If the foil subtends a fraction of the cross sectional area through the trap the probably of a neutron being absorbed on the vanadium is ~ t/f, where t is the time it takes the neutron to cross the trap. For our trap t is ~ 0.25 sec, so a value of $f = 0.1$ gives an absorption time of a few seconds rather the ~ 100 seconds typical in previous UCN lifetime experiments.

We have tested this method by activating a foil with UCN transmitted through an aluminum window separating source vacuum from atmospheric pressure.[5] The foil was activated for several minutes and then was transported by hand to a counting system that consisted of two thin plastic scintillators for detecting β and a $10 \times 10 \times 30$ cm^3 CsI(nat) detector for measuring γ-rays.

The data from this simple experiment are shown in Fig. 5 and inferred efficiencies of the detectors are given in Table 1. We believe that β particles stopped in the 250 mm thick vanadium foil and in the 250 mm thick scintillator covering account for the loss of the β counter efficiency, and that the small solid angle subtended by the CsI gives a loss in efficiency. Both of these can be improved considerably in future experiments. Fifty percent counting efficiency is a reasonable goal.

In summary, we have described a new method for counting stored neutrons in a UCN trap which may reduce the systematic errors associated with phase space evolution coupled to the long counting times of previous experiments. The short counting times that can be achieved with this technique can reduce this systematic error to levels on the order of 0.1 sec.

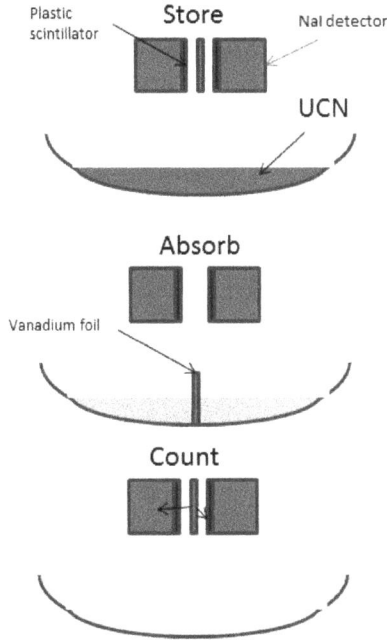

Fig. 4. Schematic of the proposed counting technique. Neutrons are stored in the trap. At the end of the storage time a vanadium foil is inserted into the trap and the neutrons are absorbed. Finally, the vanadium is raised to a position where the β and the γ from the decay of ^{52}V can be counted.

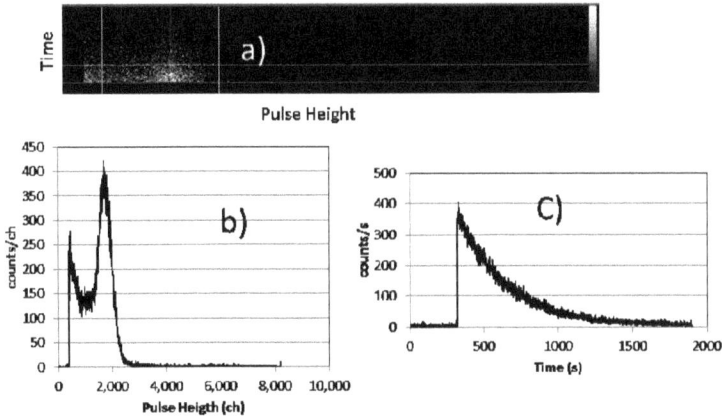

Fig. 5. Data taken to test vanadium activation for counting UCN. a) is a two dimensional plot of γ pulse height vs. time for β–γ coincidences. b) Shows the coincident γ-ray pulse height distribution summed over the region marked by the horizontal green lines. The 1.43 MeV γ-ray peak is evident. c) Shows the coincident γ-ray count rate, summed over the region marked by the vertical green lines (Color Plate 7).

Table 1. Results of the counting tests described in the text. S1 and S2 are the plastic scintillator β detectors and CsI is the γ detector. Counting rates (sec^{-1}) are shown for a 10 minute counting period from a single vanadium exposure.

	Foreground	Background	Net		
CsI singles	1542	692	850	β eff.	γ eff.
CsI $\wedge$ (S1 $\vee$ S2)	445	9	436	0.51	0.28
(S1 $\vee$ S2)	1598	42	1556		

References

1. Beringer, J., et al., *Review of particle physics*. Physical Review D, 2012. **86**(1): p. 010001.
2. Saunders, A., et al., *Performance of the Los Alamos National Laboratory spallation-driven solid-deuterium ultra-cold neutron source*. Review of Scientific Instruments, 2013. **84**(1): p. 013304-013304-10.
3. Walstrom, P.L., et al., *A magneto-gravitational trap for absolute measurement of the ultra-cold neutron lifetime*. Nuclear Instruments and Methods in Physics Research. Section A, Accelerators, Spectrometers, Detectors and Associated Equipment, 2009. **599**(1): p. 82-92.
4. Frei, A., et al., *Transmission measurements of guides for ultra-cold neutrons using UCN capture activation analysis of vanadium*. Nuclear Instruments and Methods in Physics Research Section A: Accelerators, Spectrometers, Detectors and Associated Equipment, 2010. **612**(2): p. 349-353.
5. VornDick, B., *Investigation of Vanadium foils for in situ UCN detection for the UCN $\backslash tau$ lifetime experiment*. Bulletin of the American Physical Society, 2012. **57**.

Measuring the Neutron Lifetime with Magnetically Trapped Ultracold Neutrons

H. P. MUMM*, M. G. HUBER. A. T. YUE†, A. K. THOMPSON. M. S. DEWEY

National Institute of Standards and Technology,
Gaithersburg, MD 20899, USA
* hans.mumm@nist.gov
† Previous affiliation: Institute for Research in Electronics and Applied Physics,
University of Maryland, College Park, MD 20742, USA

C. R. HUFFER, P. R. HUFFMAN, K. W. SCHELHAMMER

Physics Department, North Carolina State University,
Raleigh, NC 27699, USA

C. O'SHAUGHNESSY

University of North Carolina, Chapel Hill, NC 27599, USA

K. J. COAKLEY

National Institute of Standards and Technology,
Boulder, CO 80309, USA

We describe an experiment to measure the neutron lifetime using a technique with a set of systematic uncertainties largely different than those of previous measurements.[1] In this approach, ultracold neutrons (UCN) are produced by inelastic scattering of cold (0.89 nm) neutrons in a reservoir of superfluid ^{4}He.[2] These neutrons are then confined using a three-dimensional magnetic trap. As the trapped neutrons beta decay, the energetic electrons produced in the decay generate scintillations in the liquid He;[3,4] each decay is detectable with nearly 100 % efficiency. The neutron lifetime can be directly determined by measuring the scintillation rate as a function of time.

Keywords: Neutron lifetime; magnetic trapping

1. Introduction

Our method for measuring the neutron lifetime utilizes ultracold neutrons (UCN) confined within a three-dimensional magnetic trap. As the interaction between the neutron's magnetic moment and the trap magnetic field $\vec{B}$

conserves energy, neutrons must dissipate energy within the trapping region by other means. This occurs when 12 K neutrons (0.89 nm) downscatter in superfluid ^{4}He to near rest via single phonon emission.[2] When the neutron's spin is parallel to the magnetic field (low-field-seeking state), it will seek to minimize its potential energy by moving toward low field regions. The trap is designed such that field gradients are sufficiently small that the neutron's spin direction adiabatically follows the direction of the magnetic field. Thus, UCN with energies below the potential difference between the center and edge of the trap (trap depth) and in the low-field-seeking state (half of those downscattering from the unpolarized beam) will be trapped by the magnetic field. UCN with the opposite spin state can experience some material confinement from the material wall of the trap, but are quickly lost through upscattering.

The UCN population is thermally detached from the helium bath allowing accumulation of UCN to a density as high as $P\tau$, where P is the superthermal production rate and τ is the UCN lifetime in the trap. Neutron decay events are detected by turning off the cold neutron beam and observing the scintillation light created by the beta-decay electrons (endpoint energy of 782 keV). When an electron moves through liquid helium, it ionizes helium atoms along its track. These helium ions quickly recombine into metastable He_2^* molecules. About 24 % of the initial electron energy goes into the production of extreme ultraviolet (EUV) photons from singlet decays, corresponding to approximately 15 photons/keV.[6] These EUV photons are frequency down-converted to blue photons using the organic fluor tetraphenyl butadiene (TPB) coated onto a diffuse reflector (Gore-Tex) surrounding the trapping region. This light is transported via a series of optical elements to room temperature and detected by two photomultiplier tube (PMT)s operating in coincidence. Our detection method allows the observation of neutron decay events *in situ*, and therefore to directly measure the decay curve.

2. Apparatus

At the heart of our apparatus is a high-current Ioffe-type magnetic trap. This trap consists of an accelerator-type superconducting quadrupole magnet that provides radial confinement, combined with two aligned low-current solenoids that provide axial confinement. The quadrupole magnet, which operates at a current of up to 3,400 A, is on loan from the High Energy Accelerator Research Organization (KEK) laboratory in Japan. The two solenoids, which operate up to 250 A, were designed and constructed

Fig. 1. The neutron lifetime apparatus on the NCNR beamline (Color Plate 8).

in collaboration with American Magnetics, Inc.[7] The trap has a trapping volume of 8 l and can be operated at a depth of up to 3.1 T, although in practice a variety of depths were used for systematic and stability reasons.

A pair of 5,000 A High-Temperature Superconducting (HTS) current leads[8] on loan from FermiLab are used to bring the current from the room-temperature power supplies into the liquid helium. We designed and built a second set of HTS leads in-house for the 250 A solenoids. These two sets of leads reduce the heat input to the helium bath from a total of 8.8 W to 0.9 W as compared to conventional leads. Finally, to further reduce helium consumption and lower lead operating temperature, we incorporated a two-stage cryocooler with 1.5 W of cooling power at 4.2 K into the apparatus. This trap assembly was operated in a custom cryostat (Fig. 1) and reached > 85 % of full current before the first quench. However, due to the need to ramp the magnets on short timescales relative to the neutron lifetime, and the subsequent magnet instabilities this introduced, most data was taken at approximately 70% of design depth.

On the beam-entrance end (Fig. 1, left side), a vertical tower houses an Oxford 400 dilution refrigerator used for cooling the helium-filled cell. The second tower near the light collection system (Fig. 1, right side) houses the current leads for the trap, supplies liquid helium for the magnet and contains the thermometry, cryogen monitoring, and magnet voltage taps.

Cold (0.89 nm) neutrons enter the trapping region through a series of Teflon (vacuum) and beryllium (thermal) windows on the dewar.[9–11] Interlocking tubes of boron nitride (BN) surround the beam and shield the dewar from scattered neutrons to minimize backgrounds from neutron-induced activation. Near the front window of the helium cell the BN is coated with graphite to reduce light scattering. Neutrons scattered out of the beam are absorbed by high-purity BN shielding exterior to the light detection system while the remaining beam is absorbed at the end of the cell by the acrylic light guide. Neutrons absorbed in the acrylic do not cause significant color center formation at our present beam intensity and hence do not degrade the light detection efficiency. Thin carbon tubes line the inside of the BN to absorb any neutron-induced luminescence from color center formation in the BN. The nearly isotopically pure ^{4}He, acrylic light guides, and shielding assembly are housed in a stainless steel tube that extends through the length of the magnet. This cell is cooled to $T <$ 300 mK to reduce phonon upscattering. To minimize eddy-current heating, the refrigerator is thermally connected to the trapping region (cell) using a superfluid helium heat link as opposed to copper links. The 26 kg cell is thermally isolated using 6 Zylon support fibers.

In principle, trapped neutrons travel in the helium undisturbed until they beta-decay. Each decay is detectable with very high efficiency (depending on primarily on the low-energy threshold). The light detection system is shown in Fig. 2 and consists of a 9 cm outer diameter Goretex tube with a thin layer of TPB evaporated onto the inner surface. The end of the tube is viewed by a 9 cm diameter acrylic rod that transports the light outside the magnet bore. The light exits the low temperature region through a 5 mm thick acrylic window at the end of the trapping cell, passes through both a 6 mm thick acrylic window and a 1.5 mm quartz window at 4 K, and then into a 11.4 cm diameter acrylic rod that transports the light outside of the apparatus. Once exiting the vacuum region, light is collected via a non-imaging Winston cone and divided using a Y-shaped acrylic splitter to couple the light into two 12.7 cm diameter PMTs that are operated in coincidence mode to reduce uncorrelated low-temperature luminescence, dark-current noise and helium after-pulsing.

Pulses are digitized using transient waveform digitizers and stored for off-line analysis. This allows one to vary discrimination thresholds as well as perform pulse-shape discrimination (PSD). Background events arising from cosmic rays passing through either the acrylic or helium are tagged with approximately 30% efficiency using an array of scintillation paddles. Radiation from instruments in the surrounding area also introduce background events. The apparatus is shielded on the neutron entrance and reactor (south) sides by walls consisting of 10 cm of lead, 5 cm of polyethylene,

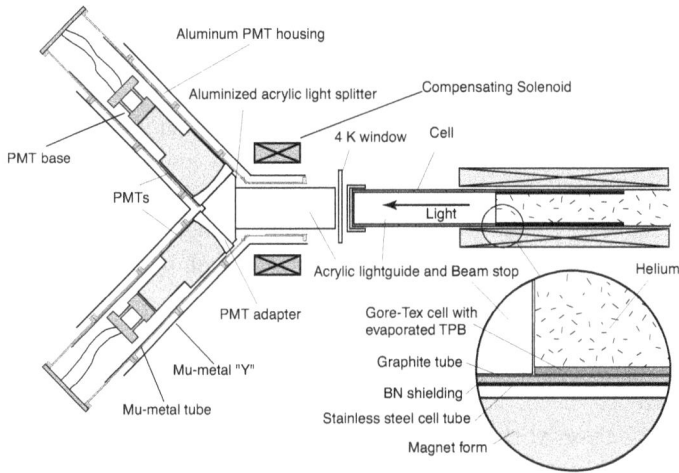

Fig. 2.   Light collection system.

and Boroflex sheets and concrete blocks on the north side . The remaining backgrounds arise from neutron-induced activation of materials within the detection region, neutron-induced luminescence, and gamma-rays Compton scattering within the acrylic.

3. Systematic Effects

In the following discussion we provide an overview of our three largest potential systematic effects and our knowledge of each. However, we emphasize that continued studies of systematic effects are underway and in most cases we describe ongoing work.

3.1. *Marginally Trapped Neutrons*

Neutron loss during the measurement process, resulting in a systematic shift in the measured lifetime, is a reoccurring problem in UCN experiments. In our apparatus, UCN with energies both above and below the trap threshold are produced during the trap loading stage. The vast majority of *above-threshold* neutrons are absorbed or up-scattered by the materials of the trap wall before data collection begins, and therefore do not affect the measurement. A small fraction, however, remain in the trap for finite times due to a combination of two mechanisms: marginally trapped quasi-stable orbits and material bottling. Here we present a brief description of our work to understand these effects, more detailed discussions can be found elsewhere.[10,11]

In support of earlier versions of the apparatus we developed a simplified 3-D analytical model for neutron loss. With this simple model, we were able to show that briefly ramping the trap to a lower field could remove a fraction of above threshold neutrons from the trap. Significantly, this analysis suggested that with a higher-field magnet, field ramping should result in considerably more efficient purging of above threshold neutrons than in shallower traps.

The development of a complete numerical simulation of neutron trajectories and wall interactions is now underway, however it is complicated by the occurrence of chaotic scattering. Although the trajectories of above-threshold neutron with escape times longer than 5 - 10 s in general can not be precisely tracked, we assume the mean survival time of an ensemble of neutrons is well determined at any time of interest.[12] Thus, simulation allows us to study the coupling between and axial and radial components of the trajectory, and estimate the time-dependent survival probability of above threshold UCN.

The previous trap, due to its shallow depth, exhibited strong material bottling. For the new KEK trap however, because the above-threshold neutrons gain more kinetic energy during a ramp, they have a higher probability to penetrate the wall potential, reducing this effect. From simulations, we can then find the ramping conditions under which we can purge practically all UCN and achieve a negligible systematic error.[10]

As a way of benchmarking these simulations we have focused on determining the systematic error in the lifetime estimate due to above threshold UCN for the KEK trap at 70 % maximum strength for two bounding material wall potentials in cases where the magnetic field is static (i.e. the field is not ramped). In one case, the cylindrical surface wall loss probability is predicted according to the multilayer potential associated with the various wall materials as seen in Fig. 3.[13] In the second case, an upper bound on incoherent scattering is added. Although these models predict very different wall loss probabilities, the systematic corrections associated with the two models are similar, $\Delta_\tau = -30.1 \pm 3.3$ s and $\Delta_\tau = -30.0 \pm 2.7$ s. We note that this is the estimated shift in the trap lifetime with no field ramping. Initial simulations that take into account ramping the quadrupole field from the initial 70 % field strength to 50 % reduce this systematic effect to less than 3 s. In future measurements, we anticipate ramping to lower field values will further ameliorate the effect.

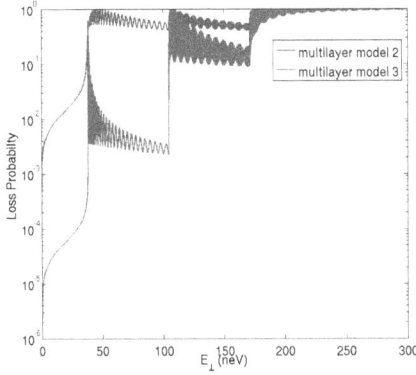

Fig. 3. The multi-layer models for the material wall potential used in modelling above-threshold neutrons (Color Plate 8).

Through a combination of analytic and numerical models, we are developing an understanding of systematics related to above threshold neutrons systematic and are refining our understanding of the conditions under which the probability of retaining significant of these neutrons is negligible. In addition, this systematic can also be experimentally tested by taking lifetime data at different minimum values of the field ramp. With the higher counting statistics afforded by the KEK magnet, one can experimentally quantify the field ramping technique and these measurements as benchmarks to further refined Monte Carlo models.

3.2. ^{3}He Purity

The isotopic ratio of natural helium $R_{34} = ^3$He$/^4$He is 1.3×10^{-6} for atmospheric helium and $(1-2) \times 10^{-7}$ for commercial helium extracted from natural gas wells.[14] For the lifetime experiment, it is essential to have significantly increased isotopic purity; the UCN loss rate Γ_{abs} due to absorption by ^{3}He is $\Gamma_{abs} = nR_{34}\sigma_{th}v_{th}$, where $n = 2.17 \times 10^{22}$ cm^{-3} is the number density of helium atoms, $\sigma_{th} = 5330$ b is the thermal neutron capture cross section, and $v_{th} = 2200$ m/s is the thermal neutron velocity. Thus a purity of $R_{34} < 10^{-15}$ is required to reduce the correction from the ^{3}He absorption to $< 2 \times 10^{-5}\tau_n$.

This purity has been demonstrated previously in 1978 by McClintock *et al.*[15] where, using helium produced using the heat flush technique from the same apparatus that produced our helium, an isotopic purity of $< 10^{-15}$ was indirectly measured using Accelerator Mass Spectroscopy (AMS) after reverse concentration. However, in order to demonstrate the required purity with a direct measurement, we have been collaborating with the AMS group at the Argonne Tandem Linac Accelerator System (ATLAS) facility at Argonne National Laboratory (ANL). We believe that the results from our latest series of runs show that the facility will be capable of making AMS measurements with sufficient sensitivity for our needs; isotopic ratio measurements with AMS can typically reach a maximum sensitivity of $10^{-15} - 10^{-16}$.

The advantage of AMS is that by accelerating ^{3}He to approximately MeV energies and passing them first through a stripper foil and then into the split pole spectrometer at ATLAS, the ^{3}He^{2+} ion peak can be unambiguously separated from background ions and molecular states of other species. The problem, then, becomes one of eliminating all possible sources of natural helium contamination from the sample gas and operating a very weak ion beam for extended periods.

Early work consistently underscored the problem of helium backgrounds. For example, a measurement carried out with a pure hydrogen plasma yielded a measurable mass-4 current and an easily detectable ^{3}He^{2+} count rate. The measured abundance was $\sim 10^{-8}$, consistent with natural contamination sources. When this background was scaled appropriately to the isotopically pure measurements, a background abundance of $\sim 10^{-12}$ would be implied. In an effort to eliminate background helium we have now built a new RF source that operates at high pressure and with good stability. The walls of the source are constructed from low helium diffusivity GE-180 and Uranium glass which is directly joined the the metal walls. All other source seals are metal. A gas handling system was constructed using two motor-controlled precision leak valves to allow gas samples to be changed without disturbing the operation of the source. In particular, this allows switching from pure hydrogen (used as both a guide beam and a test of contamination) to helium samples. With the new setup, we demonstrated good ^{3}H^{1+} production and beam stability. In addition, we produced a series of reference samples using natural helium measured to 1%. Measurements of these samples show good agreement with expectations. Finally we measured a set of ultra-pure samples from both the apparatus as well as a small quantity of gas from the original purification. Data analysis is still underway, however both samples show contamination at the few $\times 10^{-12}$ level. For reference, a purity of 5×10^{-12} would lead to a shift in the measured lifetime of 90 s.

As a result of uncertainties associated with purity issues as well as the loss of a significant fraction of our ultra-pure helium during warm-up after the last data collection run, we are in the process of constructing our own helium purifier based on the design of the McClintock system, with the important distinction that nearly all components in the system on the ultra-pure side are now metal. This apparatus is in the final testing stages. The helium produced using the new purifier will be included in our upcoming purity measurements at ATLAS.

In support of further suppressing the effects of helium contamination, we also modeled, using a three-dimensional Navier-Stokes formalism, the transport of ^{3}He in ^{4}He at 400 mK via the heat flush technique. At this temperature, our dilution refrigerator has a cooling power of $\sim$ 6 mW. We have shown theoretically that by using an easily implemented smaller diameter tube to connect the measurement cell to the mixing chamber, one can reduce the ^{3}He concentration in the measurement cells by more than two orders of magnitude (with our present geometry, we can reduce the concentration in the cell by a factor of ten). In addition, at these low temperatures, the vapor pressure of ^{3}He is considerably larger than that of ^{4}He, allowing us to preferentially pump away any remaining ^{3}He that is moved to the mixing chamber region.

With multiple approaches available and the proven purification technique,[15] we are confident that contamination of ^{3}He will ultimately not be a limiting factor in our lifetime measurement.

3.3. Backgrounds

Our detection system is sensitive to any mechanism that produces light. Thus, scintillation in the helium due to energetic charged particles as well as Compton scattering in the acrylic lightguides are potential backgrounds. These events arise from ambient radiation as well as both neutron-induced activation and luminescence. The ambient radiation gives rise to an overall constant background rate, while neutron-induced backgrounds are time dependent; some with timescales similar to the neutron lifetime. We now discuss our approach to reducing these backgrounds. All data presented here are preliminary. All figures show 20 pairs of trapping and non-trapping runs which corresponds to approximately four days of data collection.

Events are digitized and a timestamp is recorded for each. Corrections are made to the data for deadtime arising during readout of the digitizing cards and due to the hardware event veto. Corrections are applied to the pulse area and pulse height in response to the gain monitoring system. Several cuts are applied to remove background events. Initially, our electronics

(a)

(b)

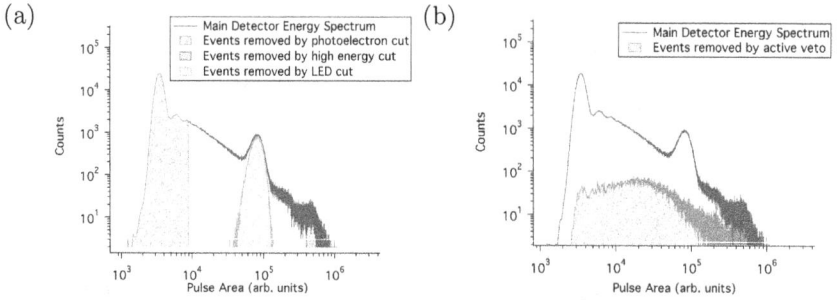

Fig. 4. (a) Application of threshold and pulser event cuts. (b) Removal of cosmic ray events (Color Plate 8).

require a coincidence between events in both photomultiplier tubes of the primary detector, thus all digitized data have events in both primary detector channels that pass a low-level voltage discrimination threshold. Events that are coincident with a separate detector that is part of the gain stabilization light pulser system are removed as can be seen in Fig. 4 (labeled as LED cut). This detector resides approximately 2 m from the apparatus and is shielded from any changes in magnetic fields generated by the apparatus. Cosmic ray events detected in external scintillators are then used to veto coincident events in the primary detector channels. The placement of these scintillators can be seen in Fig. 1. Figure 4(b) shows the spectrum of events removed in this process. More restrictive upper and lower level software cuts are then applied to the pulse area to remove uncorrelated events related to neutron induced luminescence and pulse-shape distortion due to limited digitizer dynamic range respectively. These cuts are shown in Fig. 4(a). The lower level threshold was set to the equivalent of 12 photoelectrons between the two photomultiplier tubes. Work is still in progress to determine the optimum value for the upper-level threshold; we have used conservative cuts in the following discussion.

The resulting data are then analyzed in terms of pulse shape. The kurtosis - a gaussian parameter related to the fourth moment of the pulse shape - is used to separate different classes of events. One can observe in Fig. 5(a) two distinct peaks in the data. These correspond to events that occur in either the helium itself (lower value of kurtosis) or in the acrylic light guides (higher value of kurtosis). A threshold cut is made as indicated in the figure to remove events originating in the light guides. One can clearly see this separation of events when the same data is plotted as pulse area vs. pulse height; see Fig. 5(b). Events in blue are ones unaffected by the pulse

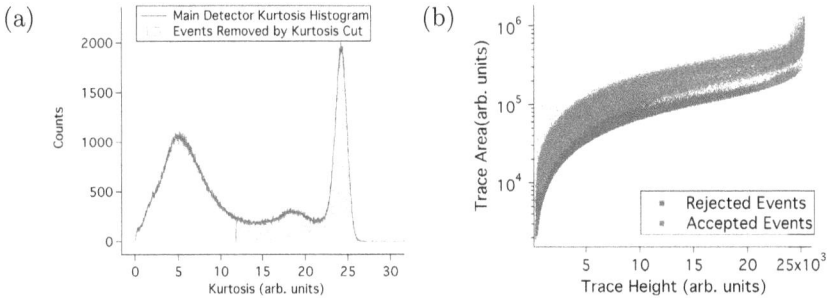

Fig. 5. (a) Events histogramed by gaussian kurtosis and illustrating pulse shape analysis of the data. (b) Two-dimensional plot of pulse area vs height; events in the lower band (associated with the plastic) not removed by the pulse shape cut are removed later with specific pulse height cuts (Color Plate 9).

shape cut, whereas red events correspond to events removed (i.e. having originated in the acrylic lightguides).

The resulting data after all corrections and all cuts have been applied is shown in Fig. 6. We expect that the mean number of photoelectrons for a neutron decay signal to be approximately 40 photoelectrons. In these data, the ambient radiation gives rise to an overall constant background event rate of order 70 s^{-1}. The neutron activation backgrounds yield time-dependent initial rates of order 10 s^{-1}. Since these are comparable in scale to the observed initial neutron decay count rate, 10 s^{-1}, an additional background subtraction procedure is required.

To accomplish this, two kinds of runs are performed; neutron trapping runs in which the magnetic field remains energized during both the trapping and observation periods, and background only runs where the field is energized only during the observation period and no neutrons are trapped during the loading phase. Data for each type of run is shown in Fig. 7.

One extracts the neutron lifetime by taking the difference between the data from the trapping and non-trapping runs. The result of this subtraction is shown in Fig. 7. In principle, one then directly fits this data to an exponential decay to extract the neutron lifetime.

Systematic effects related to imperfect background subtraction were previously investigated with data taken in the smaller apparatus when the isotopically-pure helium in the experimental cell was replaced with natural isotopic abundance helium. The resulting concentration of ^{3}He ($\approx 10^{-7}$) has a minimal effect on the neutron beam, and thus backgrounds, but the UCN trap-lifetime is shortened to less than 1 s as UCN are efficiently captured.

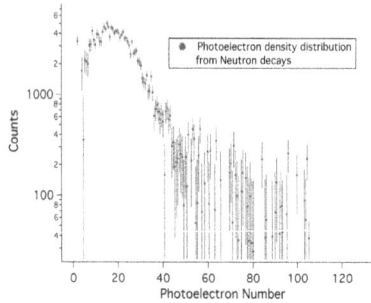

Fig. 6. Neutron trapping data after all background cuts from above (single channel) (Color Plate 9).

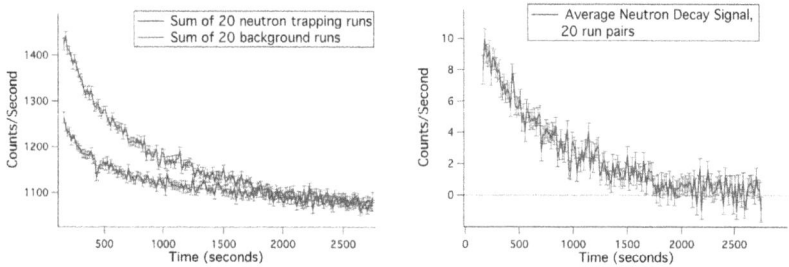

Fig. 7. Neutron trapping and non-trapping data after all background cuts (left) and the neutron decay signal after subtraction of trapping and non-trapping data runs (right) (Color Plate 9).

The data taken were consistent with zero, thus providing confidence that the decay signals observed from the low-temperature runs with isotopically pure helium originate not from imperfect background subtraction, but from trapped neutrons. Similarly, one can raise the temperature in the trap to a point where thermal up-scattering reduces the trap lifetime to well below a second. Analysis of these data is underway, but preliminary results also support the efficacy of the background subtraction technique.

4. Conclusions

The neutron decay curves extracted from 300 mK data collected using our new apparatus exhibit a trap lifetime that has a significant shift due to some, as yet, unexplained systematic effect that suggests that there exist neutron trap loss mechanisms other than beta decay. Runs where the magnetic fields are ramped to remove marginally trapped neutrons have not been able to explain this discrepancy. However as discussed above, based on recent results from the AMS measurements on the ultra-pure helium produced by McClintock, we believe that losses due to the presence of small amounts of ^{3}He in our isotopically purified ^{4}He is a likely candidate.

A significant benefit of the new apparatus is that the increased count-rates give us time to perform systematic checks such as measuring the neutron loss due to phonon up-scattering, the elimination of above threshold neutrons, and a careful characterization of imperfect background subtraction. Going forward, we intend to purify a new sample of ultrapure helium, and, using this sample continue with our systematic studies of marginally trapped neutrons.

Assuming our apparatus operates as it has previously, and taking advantage of several opportunities for improvement, we anticipate that we will be able to perform a 0.5 % lifetime measurement in 18 days, or equivalently, a measurement with a statistical uncertainty corresponding to < 3 s in a 40 day reactor cycle. Such a measurement would yet play an important role in clarifying the current uncertainty surrounding the neutron lifetime as the systematic uncertainties are very different than those of other experiments.

References

1. J. M. Doyle and S. K. Lamoreaux, *Europhysics Letters* **26**, 253 (1994).
2. R. Golub and J. M. Pendlebury, *Physics Letters* **53A**, 133 (1975).
3. D. N. McKinsey, et al., *Physical Review A* **67**, 062716 (2003).
4. D. N. McKinsey, et al., *Physical Review A* **59**, 200 (1999).
5. S. Arzumanov *et al.*, *Phys. Lett. B* **483**, 15 (2000).
6. J. S. Adams et al., *J. of Low Temp. Phys*, **13**:1121, (1998)
7. Certain trade names and company products are mentioned in the text or identified in illustrations in order to adequately specify the experimental procedure and equipment used. In no case does such identification imply recommendation or endorsement by the National Institute of Standards and Technology, nor does it imply that the products are necessarily the best available for the purpose.
8. J. M. Lock, *Cryogenics* **9** 438, (1969).
9. J. S. Butterworth *et al.*, *Rev. Sci. Instrum.* **69**, 3998 (1998).
10. L. Yang, PhD thesis, Harvard University, 2006.
11. C. M. Oshaughnessy, PhD thesis, North Carolina State University, 2010.
12. K. J. Coakley *et al.*, *J. Res. Natl. Inst. Stand. Techol.* **110** 367, (2005).
13. *Ultra-Cold Neutrons*, R. Golub, D. Richardson and S. K. Lamoreaux, OP Publishing Ltd., (1991)
14. F. Pobell. *Matter and Methods at Low Temperatures*. 2nd edition. Springer-Verlag, Berlin, 1996.
15. P. V. E. McClintock, *Cryogenics* **18** 201, (1978).

UCNτ: Study of Lifetime Measurement in a Magneto-Gravitational Trap

ALEXANDER SAUNDERS

*Los Alamos National Lab
Los Alamos, NM, 87545, USA*

D. SALVAT*, E. ADAMEK*, D. BOWMAN†, S. CLAYTON‡, C. CUDE*, W. FOX*,
G. HOGAN‡, K. HICKERSON§, A. T. HOLLEY*, C.-Y. LIU*, M. MAKELA‡,
G. MANUS*, C. MORRIS‡, S. PENTTILA†, J. RAMSEY‡, S. SAWTELLE*,
K. SOLBERG*, J. VANDERWERP*, B. VORNDICK¶, P. WALSTROM‡,
Z. WANG‡, A. R. YOUNG¶

*Indiana University, Bloomington, Indiana, USA
†Oak Ridge National Lab, Oak Ridge, TN, USA
‡Los Alamos National Lab, Los Alamos, NM, USA
§California Institute of Technology, Pasadena, CA, USA
¶North Carolina State University, Raleigh, NC, USA*

The UCNτ project is intended to develop a new measurement of the neutron lifetime using ultra-cold neutrons (UCNs) stored in a magneto-gravitational trap. In this article, we will describe the development of the experiment so far, including the trap itself, the UCN transport and monitoring system, the neutron detection methods, and the Monte Carlo simulations that have been used to model these elements. Finally, we will describe the first systematic effects that we plan to study using this apparatus.

1. Introduction

Because recent measurements of the neutron lifetime[1–4] have disagreed significantly with the previously accepted world average,[5] we have begun the development of a new experiment to measure the neutron lifetime, using ultra-cold neutrons (UCNs) stored in an asymmetric magneto-gravitational trap.[6] The neutrons are filled into the trap by a guide system including a 6 T polarizing magnet, an Adiabatic-Fast Passage spin flipper, and a super-barrier neutron detector that counts neutrons that are too energetic to be stored in the 50 cm deep, 0.67 m^3 trap. After entering the evacuated trap, the UCNs are stored using the magnetic field from an array of permanent magnets on the bottom and sides, and gravity on the top; therefore, the neutrons do not interact with any materials during the storage time. After being stored for times comparable to the neutron lifetime, the surviving UCNs are counted, either by draining them into a standard UCN detector[7] or by absorbing them on a natural vanadium foil lowered into the trap, whose neutron-induced activation is measured by detecting coincident beta particles and gamma rays from ^{52}V decay.

The main intended use of this apparatus is to enable the intensive study of the systematic effects on an experiment of this type. The systematics we intend to study fall into several broad categories: the effects of the neutron population phase space and its evolution during the storage time; UCN spectral effects on input monitor accuracy and detector efficiency; unexpected sources of UCN loss during the storage time; and detector effects, such as overall efficiency and position dependence. Our strategy to study these systematic effects is to both model them with Monte Carlo codes and directly measure each effect; the trap's ability to make storage time measurements at the statistical ~1 s level in a single day will facilitate the exploration of these systematics.

At the time of this workshop (November 2012), construction of the experiment is nearly complete, with first data using ultra-cold neutrons from the Los Alamos UCN source[8] expected to be acquired in the first months of 2013.

2. Experimental Layout

The layout of the experiment is shown in Fig. 1. The UCNs are supplied by the Los Alamos UCN source, located off the picture to the right, then polarized by a 6 T magnetic field. Since only high field seeking neutrons can be stored by the magneto-gravitational trap, the low field seeking neutrons that pass through the polarizing magnet are spin-flipped by an Adiabatic Fast Passage spin flipper, also not shown in the figure. Thus, the UCNs entering the right side of Fig. 1 are polarized to be high field seekers.

After entering the apparatus, the neutrons either enter the trap through a trap door in the bottom of the trap, or proceed past the trap door to a switcher leading to a super-barrier UCN flux monitor. Neutrons with sufficient energy to rise over the gravitational barrier presented by the angled guide, which also have enough energy to escape over the top rim of the trap, are detected in the UCN flux monitor. Thus, the monitor provides a continuous measurement of the neutron flux entering the apparatus, without depleting the population of neutrons with low enough energy to be stored in the trap.

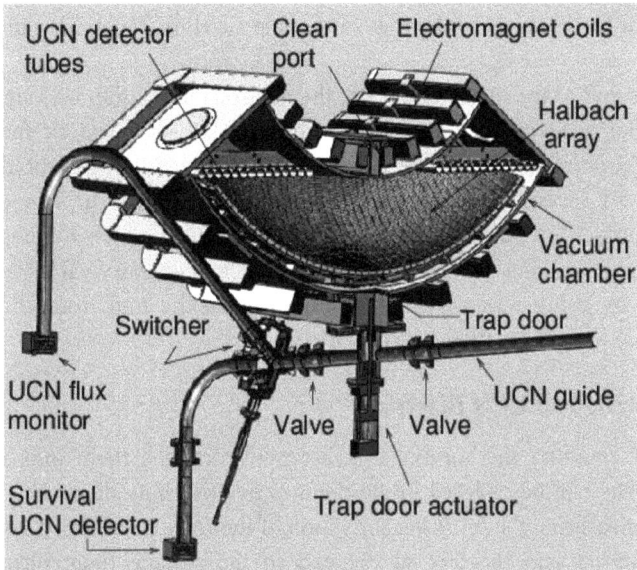

Fig. 1. Layout of the UCNτ neutron lifetime experiment (Color Plate 10).

While the higher energy neutrons are being counted by the flux monitor, lower energy neutrons are filling the trap. Once in the trap, they are confined by gravity on the top and a magnetic field provided by a Halbach array of ~1 T permanent magnets on the sides and bottom. The depth of the trap is 50 cm, and the total volume is 670 liters. The shape of the trap is that of sections of two joined tori, both with a major radius of 1.5 m and minor radii of 0.5 m and 1 m. Thus, the trap is asymmetric, with a tighter radius of curvature on one side than the other. While in the trap, the polarization of the neutrons is maintained by an externally-applied magnetic field, which is generated by coils in such a way that it is everywhere perpendicular to the field generated by the Halbach permanent magnets.

After a filling time, expected to be about 120 s, the trap door (consisting of additional Halbach magnets) in the bottom of the trap is closed to seal the magnetic trap. At this point, a cleaner, consisting of a horizontal sheet of polyethylene (not shown in Fig. 1), can be lowered into the top of the trap to absorb or upscatter any neutrons with sufficient energy to rise to its height. Neutrons upscattered by the cleaner (or any other unexpected source of loss in the trap) can then be detected by a set of 3He-loaded drift tubes mounted above the top rim of the trap, providing a continuous record of upscattering losses of neutrons in the trap. After a cleaning period, expected to be 20 s or less, the cleaner can be lifted out of the trap so that the surviving neutrons can no longer reach it.

At the end of the cleaning period, the neutrons in the trap no longer contact any material surfaces: they are confined by the Halbach magnetic fields on the sides and bottom, and by gravity on the top. They can thus be stored for up to thousands of seconds, with potentially negligible sources of loss other than neutron decay. The surviving neutrons can be counted, either by reopening the trap door and draining them through the switcher into a survival UCN detector, or in situ by using a vanadium foil lowered into the trap from above, to be described in the next section.

3. Vanadium Foil UCN Detector

Instead of counting the surviving neutrons by draining them into a standard detector, they can be counted inside the trap by absorbing them on a vanadium foil, then measuring the resulting activation of the foil. To absorb the UCNs, the foil is lowered into the trap at the end of the storage time; then, after an absorbing period, expected to be about 20 s for the initial version of the detector, the foil is extracted into an activation measurement assembly outside the trap. Since vanadium has a small but negative Fermi potential for UCNs and has a 5 barn thermal absorption cross section,[9] UCNs which impact the foil are likely to be absorbed on the 99.75% abundant 51V, resulting in a population of 52V proportional to the number of neutrons in the trap when the foil was introduced.

Vanadium-52 has a half-life of 3.7 minutes, decaying nearly 100% of the time via beta emission with a 2.5 MeV endpoint energy in coincidence with a 1.4 MeV gamma ray.[10] By detecting the beta and gamma in coincidence, background rates in the vanadium decay detector should be dramatically reduced. A schematic of the vanadium decay counter, with the foil extracted from the UCN trap for counting, is shown in Fig. 2. The beta particles are detected by a pair of scintillators lying on either side of the vanadium foil. By using two scintillators surrounding the foil, nearly complete solid angle coverage can be achieved, and cosmic rays which penetrate both scintillators can be rejected. Simultaneously, the coincident gamma ray can be detected by a pair of sodium iodide crystals.

Because the neutrons are absorbed in situ in a matter of seconds after the foil is inserted into the trap, while the resulting vanadium-52 decays with a half life of 3.7 minutes, the neutrons can be effectively counted at a much higher rate than would be otherwise expected, with reduced risk of pile up in the detector system. The initial version of the vanadium foil, designed only for testing the concept of this detector technology, has a surface area of 381 cm^2. Since the area of the mid-plane of the neutron trap is approximately 5100 cm^2, the initial version of the vanadium foil absorbs the neutrons in the trap much more slowly than is theoretically possible by using a maximal area foil completely covering the mid-plane of the trap. The ultimate version, to be designed with the aid of the Monte Carlo simulations described in the next section, should reduce the absorption time to approximately the time it takes the neutrons to travel the length of the trap, or about 1 second.

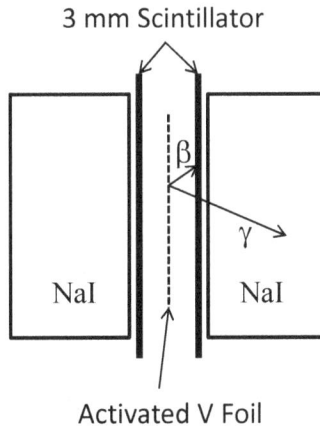

Fig. 2. Schematic of the vanadium foil activation counter. Beta particles from vanadium-52 decay are detected in a pair of plastic scintillators while the gamma rays are detected in sodium iodide crystals. Coincidence between a single scintillator and sodium iodide crystal identifies a vanadium-52 decay event.

4. UCN Tracking Simulations

In order to simulate the trajectories and interactions of UCN in the experiment, two general processes must be simulated: 1) UCN transport from the source and guides to the trap and monitor detectors, and 2) tracking of UCN in the storage trap. Because solving the equations of motion for UCN trajectories in 2) requires detailed spin tracking, while material interactions are important in 1), it is beneficial to use different sets of tools to model these two portions of the experiment.

The tracking of neutrons within the guide system uses the GEANT4UCN code.[11] This permits a detailed representation of the guide geometries that could be used in the experiment and can be used to optimize transport efficiencies. Further, experimental characterizations of the UCN source can be used to make data-driven predictions.[8] As an example, UCN are created upstream of the experiment with an experimentally determined initial velocity spectrum. From the time of flight of the UCN, we can predict the loading time of the trap for different guide configurations, and source fluctuations can be included.

For neutron tracking within the trap, the classical trajectories and various absorbing surfaces, such as the cleaning surface and vanadium foil, can be included. The analytical potential models from Ref. 6 are integrated using a 4th order symplectic integrator.[12]

The simulation of UCN in the trap can be used to investigate, for example, the time dependence of the occupied phase space of the trap, or the characteristic absorption time of UCN on the cleaner. To investigate the latter process, the trap is filled with UCNs using an initial distribution provided by the UCN guide simulations. During a short time window at the beginning of the simulation, a horizontal absorber is lowered 6 cm below the top of the trap. UCN that are marginally trapped are absorbed on the foil within this time, and the cleaner is raised. Upon raising the foil to the top of the trap, no subsequent UCN are lost over the next 20 seconds, as shown in Fig. 3.

Comprehensive simulations, like the examples described above, will be used to predict experimental observables and can be used to inform future design and investigate systematic effects.

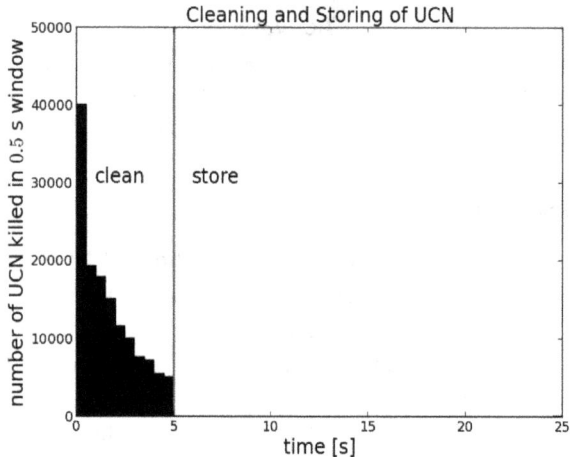

Fig. 3. Simulated cleaning of quasi-bound neutrons from the trap. The UCN source can be used to make data-driven predictions.[8] As an example, UCN are created upstream of the experiment with an experimentally determined initial velocity spectrum. From the time of flight of the UCN, we can predict the loading time of the trap for different guide configurations, and source fluctuations can be included (Color Plate 10).

5. Systematic Effects

The primary purpose of the present experimental and calculational effort is to study the systematic effects which may be present in a neutron lifetime experiment using a magneto-gravitational storage trap. Here we briefly describe a partial list of those effects, which fall into three main classes: neutron population phase space effects, UCN spectral variations, and unexpected sources of UCN loss in the trap.

The first class of systematic effects is based on the evolution of the UCN phase space in the trap. There are two main contributions in this class: the impact of quasi-bound neutron orbits on the lifetime, and the influence of phase space evolution on the neutron detector efficiency. Quasi-bound orbits are neutron orbits in the trap that are energetic enough to eventually escape over the top rim of the trap, but which spend times comparable to the neutron lifetime in the trap before they escape. Monte Carlo predictions such as those shown in the previous section suggest that the asymmetric nature of our trap causes quasi-bound orbits to very rapidly be cleaned from the trap at a reasonable cleaner depth, an issue which we can study experimentally by measuring, for example, the lifetime with different cleaner depths.

The existing vanadium foil detector and the trap-door draining detector share the feature that they sample very small sections of neutron phase space. The existing vanadium foil has a surface area of only 381 cm^2, while the trap door has an open area of 232 cm^2, sampling a trap volume of about 670 liters. Thus, if the neutron phase space density evolves through the area sampled by these detectors during the lifetime of the neutron, the effective detector efficiency can change according to the evolving local neutron density. We can test our Monte Carlo predictions of the size of this effect by varying the location of the existing vanadium foil detector when it is lowered into the trap, and determining whether there is a systematic dependence of the detector efficiency on the sampling location.

Since the incoming UCN flux monitor counts the neutrons with energies too high to be stored in the trap, while the lifetime measurement uses the low energy storable neutrons, any variation in the incoming neutron velocity spectrum could affect the relative flux monitor efficiency, thus resulting in an incorrect estimate of the number of neutrons loaded into the trap for each storage measurement. Since the velocity spectrum emerging from the Los Alamos UCN source can be affected by details of the operation of the source, outside of the lifetime experiment's direct control, this effect must be carefully measured. One way to monitor this effect is to draw off and count some fraction of the storable neutrons during the filling process, and use those neutrons to calibrate the super-barrier incoming flux monitor efficiency. Also, high statistics spectral measurements can easily be accomplished between fills by continuously running the super-barrier and draining detectors while the trap door is closed, thus providing a very sensitive measure of any relevant spectral changes.

Unexpected sources of UCN loss are by their nature hard to anticipate before the trap is in operation, but there are several sources of loss that we expect to have to measure and control. These expected sources of loss include neutron spin flips, neutron loss by scattering or absorption on residual gas, heating by vibrations of the trap walls, and weak spots in the magnetic field produced by the Halbach array. Neutron spin flips can be caused by field zeroes (due perhaps to non-homogeneity of the spin holding field) or by neutrons failing to maintain the adiabaticity condition as they move through the combined holding and Halbach magnetic fields. This loss mechanism can be amplified and studied by, for example, depowering or reversing one of the ten coils which supply the holding field, thus introducing high field gradients and field zeroes into the trap.

Neutron loss on residual gas is estimated to be a very small effect for reasonably attainable vacuum levels, at the 10^{-6} Torr level. However, the effect can be studied by varying the vacuum level in the trap and by poisoning the trap with different residual gases. Vibration of the trap walls can be studied by intentionally adding vibrational energy to the trap with a shaker and measuring the amount of vibration present. Weak spots in the Halbach array, caused by broken or demagnetized permanent magnet elements, can be enhanced by intentionally adding reversing magnets or lossy materials to the trap walls.

6. Conclusion

The construction of the UCNτ lifetime apparatus is nearing completion, and first data acquisition is expected in the early months of 2013. The experiment apparatus, consisting of a magneto-gravitational neutron storage volume, with associated neutron supply system and detection systems, is intended to be used as a testbed for studying the systematic effects associated with a next generation trapped neutron lifetime experiment. Innovative features include an asymmetric, material-free storage trap and an *in situ* neutron detection scheme based on absorption of the neutrons on a vanadium foil inserted into the trap. We hope to eventually achieve a 1 second total uncertainty measurement of the neutron lifetime using the present apparatus with minimal modifications, and to perform the research and development that will lead to the design of an experiment that can reach the 0.1 second uncertainty level.

References

1. Arzumanov, S. S. *et al.* (2012). *JETPL*, **95**, p. 224.
2. Pichlmaier, A. *et al.* (2010). *Phys. Lett. B*, **693**, p. 221.
3. Serebrov, A. *et al.* (2005). *Phys. Lett. B*, **605**, p. 72.
4. Mampe, B. *et al.* (1993). *JETPL*, **57**, p. 82.
5. Beringer, J. *et al.* (2012). *Phys. Rev. D*, **86**, p. 01001.
6. Walstrom, P. L. *et al.* (2009). *Nucl. Inst. and Meth. A*, **599**, p. 248.
7. Salvat, D. *et al.* (2012). *Nucl. Inst. and Meth. A*, **691**, p. 109.
8. Saunders, A. *et al.* (2013). *Rev. Sci. Inst.*, **84**, p. 013304.
9. *Neutron News*, (1992). Vol. **3**, No. 3, pp. 29-37.
10. *ENDSF*, (1994). NDS **71**, p. 659.
11. Atchison, F. *et al.* (2005). *Nucl. Inst. and Meth. A*, **552**, pp. 513-521.
12. McLachlan, R. I. and Atela, P., (1992). *Nonlinearity*, **5**, p. 541.

A Comparison of Two Magnetic Ultra-Cold Neutron Trapping Concepts Using a Halbach-Octupole Array

K. LEUNG*,†, S. IVANOV, F. MARTIN, F. ROSENAU,

M. SIMSON, O. ZIMMER‡

Institut Laue-Langevin, 6 rue Jules Horowitz, 38042 Grenoble Cedex 9, France
* kkleung@ncsu.edu
‡ zimmer@ill.fr

This paper describes a new magnetic trap for ultra-cold neutrons (UCNs) made from a 1.2 m long Halbach-octupole array of permanent magnets with an inner bore radius of 47 mm combined with an assembly of superconducting end coils and bias field solenoid. The use of the trap in a vertical, magneto-gravitational and a horizontal setup are compared in terms of the effective volume and ability to control key systematic effects that need to be addressed in high precision neutron lifetime measurements.

Keywords: Neutron lifetime; ultra-cold neutrons; UCN; magnetic trapping

1. Introduction

The free neutron undergoes β-decay via n $\rightarrow$ p + e$^-$ + $\bar{\nu}_e$. Precise measurements of the mean lifetime τ_n are used for obtaining the universal weak coupling constants of the nucleon from which one derives important semi-leptonic weak cross sections. They are also needed for searches of beyond Standard Model physics, and for calculations of primordial helium abundance in Big Bang Nucleosynthesis. This is described in more details elsewhere in these proceedings, as well as in various review papers on the neutron particle physics field[1,2] or on τ_n specifically.[3,4]

Ultra-cold neutrons (UCNs) are free neutrons with kinetic energies less than the neutron optical potential of well-chosen materials so that they can be confined in a material "bottle" via total internal reflections. For instance, beryllium, a commonly-used material for UCN reflection, has $V_{Be} = 252$ neV, corresponding to a velocity of ≈ 7 m s^{-1}. UCNs stored in

---

†Currently at North Carolina State University, 2401 Stinson Dr., Riddick 421, Raleigh, NC 27695, USA.

bottles[a] for measuring τ_n have been used in the most precise experiments to date.[7–12]

Due to the size of the neutron's magnetic moment μ_n of $60\,\mathrm{neV\,T^{-1}}$, the Stern-Gerlach force can be used to confine UCNs also. This technique[13–15] offers a method for confining UCNs free from energy-dependent wall losses that requires to be corrected for. While the traditional *"counting the survivors"* scheme can be used for determining τ_n, extracting and detecting charged decay products is also possible with magnetic traps. A strong motivation for using magnetic trapping is to improve upon material bottle experiments with a better control and reduction of systematic corrections.[16–19]

In magnetic bottles losses of UCNs due to depolarization at the weak field regions has been shown theoretically to be suppressible if a sufficiently strong bias field is used.[18,20] However, the effects caused by phase-space evolution of the UCN gas and the issue of how to effectively remove UCNs with energy just above the trapping threshold must be carefully addressed by high-precision measurements. There is also the possibility of a gradual warming of the UCN spectrum, caused by magnetic noise or mechanical vibrations, which has largely been unexplored. With these issues in mind, two concepts using an Halbach-octupole array combined with a superconducting coil assembly are discussed in this paper.

Previously, the idea of using a UCN production volume integrated inside a horizontal, sliding magnetic bottle was presented.[21] Its goal was to extract produced UCNs to vacuum in order to avoid losses due to interactions with the superfluid helium converter as well as to avoid the dilution of the high density of UCNs offered by super-thermal production.[22] In this scheme, due to the geometry of available cold neutron beams, the long axis of the trap is required to be horizontal. However, after further exploration of the idea, combined with the recent success of a high-density superfluid helium UCN source using a vertical, window-less extraction system where transmission losses through windows are eliminated,[23] we opted to use an external UCN source for filling the trap instead.

2. Design of Magnetic Fields

The central component of the trap is the 1.2 m long 32-piece Halbach-type[24] octupole array for radial UCN confinement (shown in Fig. 1). The octupole has an inner bore radius $R = 47\,\mathrm{mm}$ and a nominal magnetic field of $B = 1.3\,\mathrm{T}$ at its surface (at room temperature). While the field

---

[a]The exceptions being in-beam measurements.[5,6]

near the center is that of an ideal octupole $B(\rho) \propto \rho^3$, deviations appear near R due to the discrete number of magnets. B from 2D finite element calculations with FEMM[25] at different azimuthal angles ϕ and distances from R is shown in Fig. 2. The weakest B for a fixed radial position ρ occurs at the off-pole pieces.[b]

Fig. 1.   The octupole magnet array assembly: (a) rods used to fix the rotation of the 12 modules, (b) split aluminum shell for holding the modules together, (c) stainless steel end plates fixed to the aluminum shell, and (d) brass pressing screws. The ribbed structure of the aluminum shell provides good thermal contact for cooling the magnets to $\sim 120\,\mathrm{K}$. An individual module is shown in Ref. 21.

These calculations also show that the magnetic material at the off-pole pieces' inner corners are the most susceptible to demagnetization. To model how this might affect B in the bore, these corners were rounded out with a 1.5 mm radius (and replaced with an air gap). This causes sharp dips right at the corners and a slight enhancement at the center of the piece (also in Fig. 2). These features become insignificant at distances $\gtrsim 2\,\mathrm{mm}$ from R. Conclusions are the same for different corner geometries (quarter-circular and chamfer cut). If chips in the brittle NdFeB material exist, then similar dips could also exist. It is therefore wise to keep UCN trajectories passing too close to these regions.

---

[b]The pieces are called off-pole when they have their magnetization vector parallel or anti-parallel to the $\hat{\phi}$ unit vector at the centre of that piece.

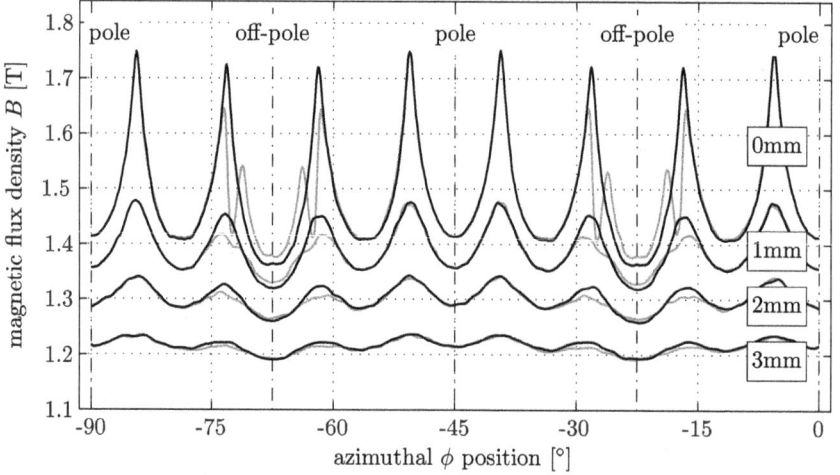

Fig. 2. 2D finite element calculations of the Halbach-octupole. B is plotted for ϕ spanning a quadrant in the xy-plane and for different distances from R (indicated by the boxed numbers). The ϕ positions of the centers of the pole and off-pole pieces are indicated. The fainter lines for each wall distance are from modeling the demagnetized magnets with rounded corners.

To understand the field at the ends of the array 3D finite element calculations were performed with RADIA.[26] This revealed that $B(R)$ drops to $< 0.8\,\mathrm{T}$ at the end of the array and there exist axial components $B_z \approx 0.8\,\mathrm{T}$ that are strongest near the surface of the pole pieces. These features are shown in Fig. 3 and have been confirmed with Hall probe measurements.[27]

A superconducting coil assembly (depicted in Fig. 4) consisting of a small end coil, a bias field solenoid, and a large end coil is used for axial UCN confinement and elimination of the low-field region. The fields that can be produced by these coils when they are run at their maximum current of 300 A are 1.7 T, 1.2 T, and 5 T,[c] respectively. The 30 cm inner diameter of the coils causes only small radial field components B_ρ to exist at R, hence there is only a small cancellation with B_ρ from the octupole array. The end coils are slotted inside the octupole array so that the region of strong B_z cancellation is situated away from the trapped UCNs. Two magnetic field configurations from the coils that will be discussed in the next section are also shown in Fig. 4. The bias field is chosen to be $> 0.1\,\mathrm{T}$ for preventing depolarization in the low field region.

---

[c]This allows focussing decay protons onto a small detector. This might be done in later experiments but will not discussed in this paper.

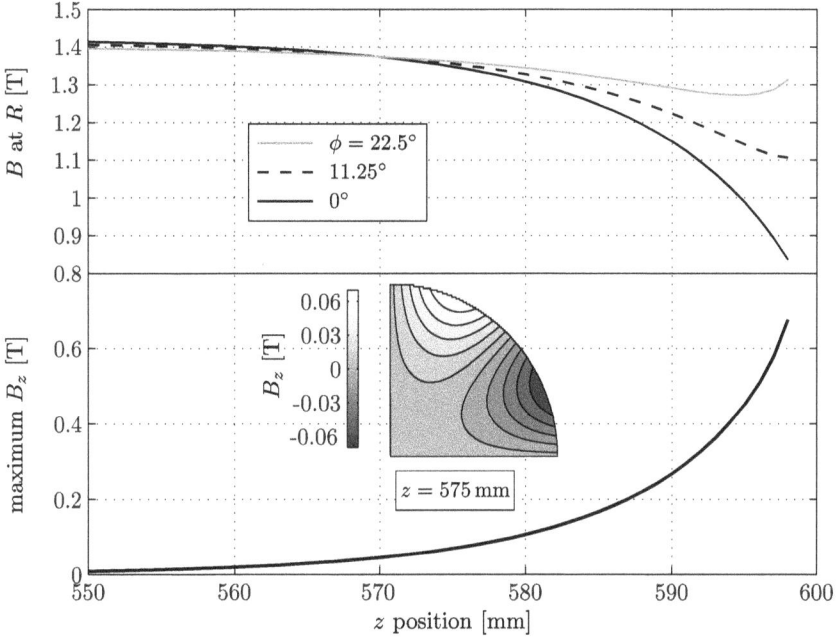

Fig. 3. The field at the ends from solely the octupole array: B at R becomes weaker and strong B_z components start to emerge. A quadrant slice of B_z at $z = 575\,\text{mm}$, the central position of the end coil placed closest to the end of the array (see Fig. 4).

Combining the 3D calculations with all field sources, it was shown that demagnetization of the permanent magnets can be avoided by cooling them to $\sim 120\,\text{K}$, which leads to a strong increase in the coercivity. This cooling also increases B at R by $\sim 10\%$, which is not included in the discussion to keep the estimates of the trap depth conservative.

3. Trapping Potential and Effective Volume

The total energy of a neutron E_n at a point $\vec{r} = (x, y, z)$ in space is given by:

$$E_n = E_{\text{kin}}(\vec{r}) + E_{\text{pot}}(\vec{r}) = E_{\text{kin}}(\vec{r}) + V_{\text{grav}}(z) + V_{\text{mag}}(\vec{r})$$

$$= E_{\text{kin}}(\vec{r}) + mgz \pm \mu_n B(\vec{r}), \tag{1}$$

where E_{kin} is the kinetic energy, m is the neutron mass, g is the gravitational acceleration, and z is the height. A constant offset is neglected to place E_{pot} at the potential minimum of the trap. In the last term, the plus sign

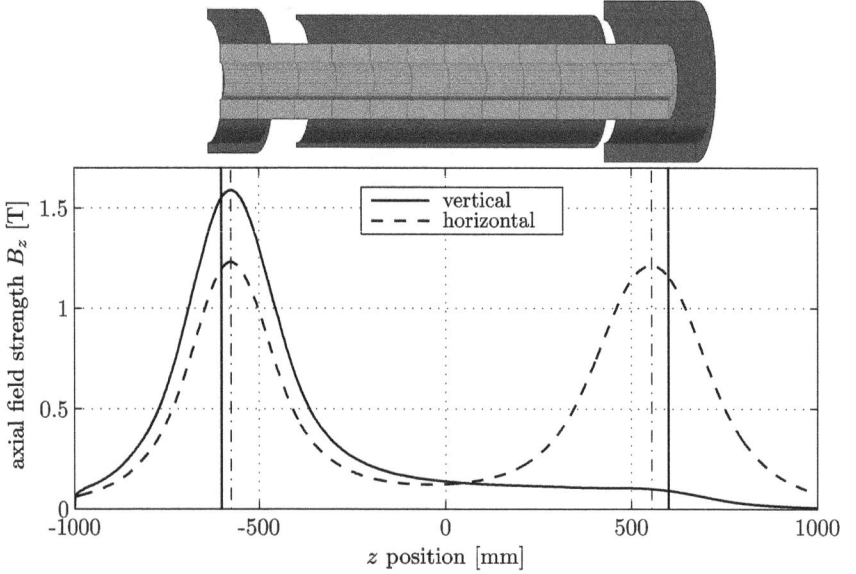

Fig. 4. *Top:* a to-scale drawing of the octupole array and the surrounding assembly of superconducting coils. *Bottom:* the axial field from the coils for the two trapping configurations. The vertical solid lines show the end of the array and dash-dotted lines the center of the end coils.

refers to the *low field seeking* spin-state. Maps of E_{pot} for a horizontal configuration, where both end coils are required for axial confinement, and for a vertical configuration, where only the bottom end coil is required due to the gravitational potential $mg = 102\,\text{neV}\,\text{m}^{-1}$ (*magneto-gravitational* trap), are shown in Figs. 5 and 6. In order to populate the trap with UCNs, current in the small end coil will be lowered temporarily to reduce E_{pot} there.

The depth of a pure magnetic trap E_{trap} is defined by the maximum E_{n} a UCN can have without the possibility of it exiting the trap at the ends or making contact with material walls. Since an inner tube will be placed inside the magnet bore, a reduction in the radius of the trap of $1.5\,\text{mm}$ is taken into account for determining E_{trap}.

Clearly, a UCN cannot explore regions in a trap where $E_{\text{n}} < E_{\text{pot}}(\vec{r})$ so the accessible volume is energy-dependent. It is useful to define the effective volume[28] $V_{\text{eff}}(E_{\text{n}})$ that can be expressed as:

$$V_{\text{eff}}(E_{\text{n}}) = \text{Re}\left[\int_V \sqrt{\frac{E_{\text{n}} - E_{\text{pot}}(\vec{r})}{E_{\text{n}}}}\,\mathrm{d}V\right], \qquad (2)$$

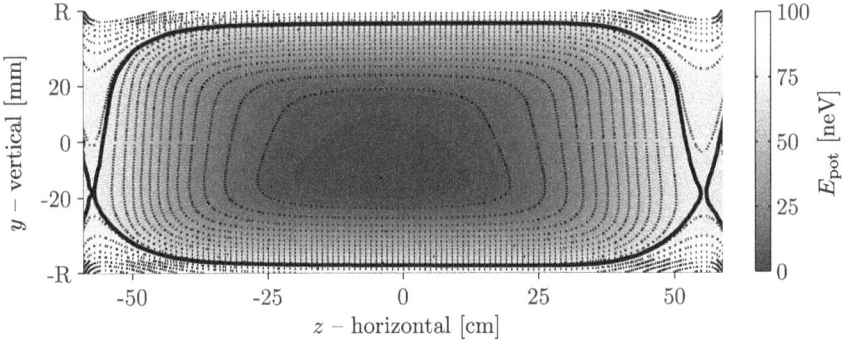

Fig. 5. A map of E_{pot} for the horizontal configuration for the $\phi = 90°$ slice aligned vertically with the axial fields shown in Fig. 4. The thick solid line shows the contour for $E_{\text{trap}} = 63\,\text{neV}$ and the dotted contour lines are placed at increments of 5 neV. The lowest E_{pot} in the trapping region is defined as 0 neV.

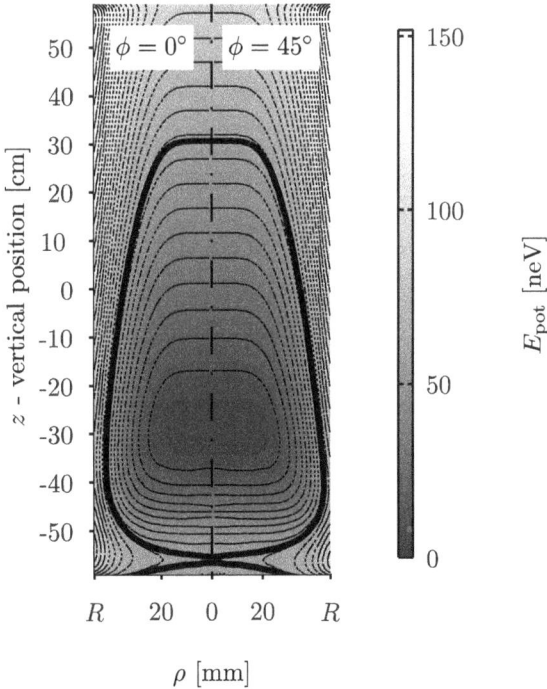

Fig. 6. A map of E_{pot} for the vertical configuration with the axial fields shown in Fig. 4 for $\phi = 0°$ (maximum B_ρ reinforcement) and $\phi = 45°$ (maximum B_ρ cancellation) slices. The thick solid line is for $E_{\text{trap}} = 48\,\text{neV}$ and the dotted contour lines are placed at increments of 5 neV. The lowest E_{pot} in the trapping region is defined as 0 neV.

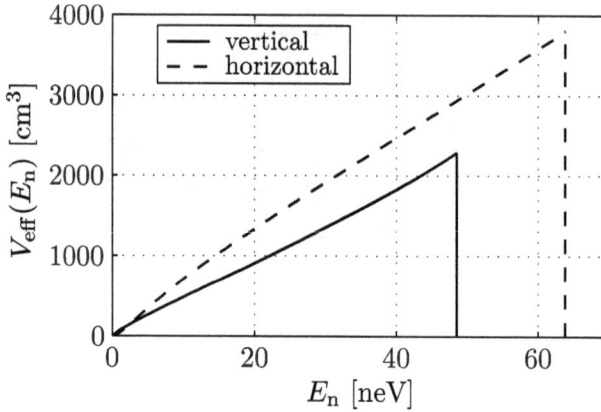

Fig. 7.   The UCN energy-dependent effective volumes for the two configurations.

where $\int_V$ is over the volume of the trap and dV is the volume element at $\vec{r}$. This is shown for the two configurations in Fig. 7. The total number of UCNs stored in the bottle is then given by: $\int_0^{E_{\text{trap}}} n(E_n) V_{\text{eff}}(E_n) dE_n$, where $n(E_n)$ is the energy-dependent spectral UCN density. If this is calculated for the typical Maxwellian spectrum, where $n(E_n) \propto \sqrt{E_n}$, then the number that can be stored in the horizontal configuration is ~ 2.5 times greater than in the vertical.

4. Advantages of the Vertical Configuration

Despite the larger number of UCNs that can be stored in the horizontal setup, there are distinct advantages for the vertical setup in terms of systematic effects.

In the horizontal configuration, the two magnetic mirrors can trap the charged decay products. *In-situ* counting with a high efficiency is possible only by extracting the protons with high-voltage electrodes ($> 5\,\text{kV}$) due to their low maximum kinetic energy of $750\,\text{eV}$. The trapped electrons can produce ions and electrons in the residual gas of the vacuum, and the presence of the electric field can accelerate them causing production of secondary charges up to electrical break down. For instance, even in the aSPECT neutron β-decay apparatus, with its excellent vacuum conditions, this effect has been problematic.[29,30] The vertical configuration avoids the electron trapping problem since the upper coil is not required for UCN confinement. Moreover, this allows electron counting as the detection scheme and thus removes the need for the high-voltage system.

A key systematic effect in high-precision measurements is poor cleaning of above-threshold UCNs from a trap. For a horizontal configuration, a scheme of ramping up and down the bias field has been used for this; however this results in a significant ($\sim 50\%$) loss of UCNs.[31] The use of a UCN reflecting paddle along the length of a trap to induce mode-mixing reflections has been suggested before.[32] However, removing the paddle from the reach of UCNs by rotation or by sliding it out of the trap causes undesired doppler heating of the UCNs after cleaning (for the latter case, heating can occur upon non-specular reflections[33]).

In the vertical, magneto-gravitational configuration, UCNs have to fall and reflect off the bottom of the trap. The insertion of a piston that induces mode-mixing reflections from the bottom can thus be used for cleaning. Furthermore, retracting it does not cause doppler heating. The details of this procedure will be published elsewhere.

Another advantage offered by this configuration is the opening at the bottom of the trap, which allows high-field seeking and above-threshold UCNs to exit the trap quickly. By monitoring these with a UCN detector during storage, it will allow systematics associated with depolarization and warming of UCNs (e.g. due to magnetic or mechanical vibrations) to be accessed experimentally, without the need to rely on theoretical calculations.

Finally, if a fill-and-empty scheme is used to determine τ_n—as will be the case for initial measurements—emptying of the trap from the bottom will have weak sensitivity to the phase-space evolution of the UCNs, a key topic discussed in these proceedings.

5. Conclusion

A detailed description of the magnetic fields from the Halbach-octupole array and the superconducting coil assembly design has been given. The trapping potential and the effective volumes for a horizontal and a vertical geometry were compared, with the former allowing the storage of a factor ~ 2.5 more UCNs. However, the advantages offered by the vertical configuration for controlling key systematic effects outweighs the reduced statistical sensitivity, which can be addressed with the further development of the new, compact superfluid helium UCN source.[23] These factors led us to the vertical, magneto-gravitational design for our planned high-precision neutron lifetime measurements.

References

1. H. Abele, *Prog. Part. Nucl. Phys.* **60**, 1 (2008).

2. D. Dubbers and M. G. Schmidt, *Rev. Mod. Phys.* **83**, 1111 (2011).
3. S. Paul, *Nucl. Instrum. Meth. A* **611**, 157 (2009).
4. F. E. Wietfeldt and G. L. Greene, *Rev. Mod. Phys.* **83**, 1173 (2011).
5. Y. Arimoto, *et al.*, *Prog. Theor. Exp. Phys.*, 02B007 (2012).
6. J. S. Nico, *et al.*, *Phys. Rev. C* **71**, 055502 (2005).
7. W. Mampe, *et al.*, *Phys. Rev. Lett.* **63**, 593 (1989).
8. W. Mampe, *et al.*, *J. Exp. Theor. Phys. Lett.* **57**, 82 (1993).
9. A. Pichlmaier, *et al.*, *Phys. Lett. B* **693**, 221 (2010).
10. S. S. Arzumanov, *et al.*, *JETP Lett.* **95**, 224 (2012).
11. A. Serebrov, *et al.*, *Phys. Lett. B* **605**, 72 (2005).
12. A. P. Serebrov, *et al.*, *Phys. Rev. C* **78**, 035505 (2008).
13. V. V. Vladimirskii, *Sov. Phys. JETP* **12**, 740 (1961).
14. W. Paul, *et al.*, *Z. Phys. C Part. Fields.* **45**, 25 (1989).
15. P. R. Huffman, *et al.*, *Nature* **403**, 62 (2000).
16. O. Zimmer, *J. Phys. G Nucl. Partic.* **26**, 67 (2000).
17. S. Materne, *et al.*, *Nucl. Instrum. Meth. A* **611**, 176 (2009).
18. P. L. Walstrom, *et al.*, *Nucl. Instrum. Meth. A* **599**, 82 (2009).
19. V. F. Ezhov, *et al.*, *Nucl. Instrum. Meth. A* **611**, 167 (2009).
20. A. Steyerl, *et al.*, *Phys. Rev. C* **86**, 065501 (2012).
21. K. K. H. Leung and O. Zimmer, *Nucl. Instrum. Meth. A* **611**, 181 (2009).
22. R. Golub and J. M. Pendlebury, *Phys. Lett. A* **53**, 133 (1975).
23. O. Zimmer, F. M. Piegsa and S. N. Ivanov, *Phys. Rev. Lett.* **107** (2011).
24. K. Halbach, *Nucl. Instrum. Meth.* **169**, 1 (1980).
25. D. C. Meeker, Finite element method magnetics FEMM, `http://www.femm.info`, (version 4.0.1, build 03 Dec. 2006).
26. O. Chubar, P. Elleaume and J. Chavanne, *J. Synchrotron. Radiat., SRI97 Conference August 1997* **5**, 481 (1998).
27. K. Fraval, Neutron life-time measurement using magnetic trapping of ultracold neutrons, Internship Report (unpublished), PHELMA, (Grenoble, France, 2009).
28. R. Golub, D. Richardson and S. K. Lamoreaux, *Ultra-Cold Neutrons* (IOP Publishing Ltd., 1991).
29. O. Zimmer, *et al.*, *Nucl. Instrum. Meth. A* **440**, 548 (2000).
30. G. Konrad, *et al.*, *Nuclear Physics A* **827**, 529c (2009).
31. S. N. Dzhosyuk, *et al.*, *J. Res. Natl. Inst. Stan.* **110**, 339 (2005).
32. J. D. Bowman and S. I. Penttila, *J. Res. Natl. Inst. Stan.* **110**, 361 (2005).
33. K. Leung, Development of a new superfluid helium ultra-cold neutron source and a new magnetic trap for neutron lifetime measurements, PhD thesis, Technische Universität München, (Garching, Germany, 2013).

Polarizing Ultra-Cold Neutrons for
the Superconducting Trap PENeLOPE

R. PICKER

TRIUMF
Vancouver, Canada
rpicker@triumf.ca

W. SCHREYER*, F. HAAS, F.J. HARTMANN, M. LOSEKAMM,

S. PAUL, R. STOEPLER, C. TIETZE

Physik Department E18, Technische Universität München
Garching, Germany
* w.schreyer@tum.de

PENeLOPE (Precision Experiment on the Neutron Lifetime Operating with Proton Extraction) is a novel experiment to measure the lifetime of the free neutron. It features magneto-gravitational storage of ultra-cold neutrons; only one spin state of the neutrons can be stored magnetically, hence a polarization system is necessary. In contrast to most other magnetic storage experiments, the magnetic field is ramped up from zero after filling, which results in a complete spatial and energetic separation of the two spin states; this allows the use of novel techniques in cleaning the trap from the unwanted spin state in addition to pre-polarization. A polarization of 99.98% should be achievable.

Keywords: Ultra-cold neutrons (UCN); neutron lifetime; magnetic trapping; polarization.

1. The Neutron Lifetime

The lifetime of the free neutron τ_n is a fundamental parameter in the Standard Model of particle physics. A precise knowledge is crucial for a better understanding of the Cabibbo-Kobayashi-Maskawa (CKM) mixing matrix:[1] if combined with measurements of the electron asymmetry A[2,3] or the neutrino asymmetry a[4] the first element V_{ud} of the CKM matrix can be computed; together with data from Kaon decay,[5] the unitary of the CKM matrix can be tested and the validity of the three generation Standard Model probed. Additionally, a better knowledge of τ_n contributes to sharpening solar models and the understanding of Big Bang nucleosynthesis.

The most recent measurements of the lifetime showed large discrepancies,[6,7] and were partially corrected much later;[8,9] hence, the Particle Data Group (PDG) decided to inflate the error bar in its latest edition with $\tau_n = 880.1 \pm 1.1$ s and calls upon experimenters for new, more precise experiments to clarify the picture.[10]

2. Measurement Techniques

The measurements included in the PDG value applied either the beam method using cold neutrons or the storage method for ultra-cold neutrons (UCN).[11] The latter mainly used material storage bottles to trap UCNs by total reflection[6] and also gravity in some cases.[7,9] Using appropriate materials like Fomblin®, long storage lifetimes[a] could be achieved, but due to unavoidable losses in material traps, extrapolation methods to zero loss had to be applied, restricting the precision reach to around 1 s, so far.

The magnetic moment of the neutron $\mu_n = -60.3$ neV/T allows to store ultra-cold neutrons magnetically in a gradient field $F_{mag} = \pm\mu_n\nabla|\mathbf{B}|$. The direction of the magnetic force depends on the orientation of the spin with respect to the magnetic field; neutrons attracted by larger flux density regions are called high-field seekers (HFS), and low-field seekers (LFS) if they are repelled by higher fields. Creating appropriate field configurations, one neutron spin species can be stored magnetically; since the flux density increases towards current-carrying wires or permanent magnets, low-field seekers are the logical choice for magnetic traps if material interactions shall be avoided. If the adiabatic condition $\omega_L \gg \dot{\mathbf{B}}/|\mathbf{B}|$ is fulfilled, UCNs may be stored virtually losslessly in contrast to material traps. This should allow to decrease the measurement error significantly in next generation experiments.

3. Polarization Techniques

When storing neutrons in magnetic traps, both species, LFS and HFS, will have very different storage lifetimes in the trap: whereas the low-field seekers of sufficiently low kinetic energy will not interact with the materials surrounding the magnetic well, the high-field seekers are accelerated towards the high fields in the vicinity of the current-carrying coils or permanent

---

[a]Notice the nomenclature used here: storage time is the preset time that the UCN shall be stored, storage lifetime the statistical lifetime due to β-decay and other UCN loss mechanisms.

magnets. This leads to material interaction, which can be total reflection, absorption or up-scattering; the two latter ones will reduce the storage lifetime of high-field seekers in the trap. As a consequence, the presence of both spin species of neutrons leads to a lower measured neutron lifetime and has to be avoided.

This can be achieved, by either only filling in correctly polarized neutrons (pre-polarization method) or eliminating high-field seekers from the trap (in-situ polarization) before starting the actual measurement.

3.1. *Pre-polarization*

Ultra-cold neutrons can be polarized relatively easily due to their low energy by installing a magnetic barrier between the UCN source and the experiment. If the magnetic potential $U = \pm\mu_n|B|$ of the barrier is higher than the kinetic energy of the UCNs, only high-field seekers will be transmitted providing 100% polarization[b] right after the barrier.[3] Since HFS cannot be stored, a 180° spin flip has to be performed on all neutrons; several techniques can be applied reaching close to unity spin flip efficiency.[12,13] The spin flipper can either be placed outside of the neutron storage volume or inside.

In the first case, the magnetic fields between the polarizer and the neutron trap have to be carefully laid out to guarantee adiabatic transport: smooth transitions between different magnitudes and orientations of the magnetic field, and no zero field regions. Most likely, this transport will involve material reflections as all guides for UCNs used up to now are material guides. Every contact with the material surface entails a certain absorption or up scattering probability, but also a non vanishing spin-flip probability. For suitable materials, these are in the region of 10^{-5} per bounce,[14,15] leading to significant depolarization, depending on the guide layout.

In the latter case, the high-field seeking UCN are filled into the trap; the spin flip is performed inside the low field region of the storage volume creating a neutron population that can be magnetically confined. Again, two scenarios can be imagined: a UCN guide is mechanically inserted into the low field region of the trap, the end of which is equipped with the spin flipper, so the neutrons become low-field seekers when they exit the guide. This has the advantage, that no material storage capabilities are necessary and the walls of the trap can be made of any material. A second option is

---

[b]Polarization is defined as $P = (\uparrow - \downarrow)/(\uparrow + \downarrow)$, where $\uparrow$ is the number of neutrons of one spin alignment and $\downarrow$ the other, respectively.

to fill the high-field seekers in, who have to be stored by necessarily neutron reflecting walls until a satisfactory UCN density is reached. Only then, spin flip coils around the storage volume can be activated, creating the low-field seeker population. The drawback here is that high-field seekers, surviving due to the non-unity spin flip efficiency, might have a much longer storage time than for the first option due to the UCN reflective walls.

3.2. In-situ Polarization

One can avoid the additional effort of building a polarizer, spin-flipper and adiabatic UCN guide system if the polarization of the neutrons can be achieved within the trap. The mechanism depends on the filling procedure of the storage volume: either the magnetic field always protects low-field seekers from hitting the material walls or during filling the walls are exposed to UCNs.

In the first case (magnetic filling), a part of the magnetic walls for the UCNs has to be lowered (most probably by turning on or off a dedicated coil[16] or physically moving permanent magnets[17]) to allow loading of low-field seekers. Once a sufficient UCN density is reached, the magnetic orifice is closed; low-field seekers will be stored magnetically and high-field seekers can be removed from the trap by coating the inner walls of the storage volume with a material that has a Fermi potential close to zero. Within seconds, a polarization close to unity can be reached. The challenge hereby is to create a large enough orifice to facilitate fast loading and unloading of the trap. Other loading techniques are possible to avoid the necessity of material storage during the filling process: A neutron lift was used for a permanent magnet trap, lowering the neutrons into the trap from above.[16] In-situ creation of the ultra-cold neutrons in a superfluid helium filled storage volume is used for a superconducting trap to avoid fast ramping.[18] A retractable filling tube can also be inserted in the trap region with a spin-flipper at the end: high-field seekers can overcome the magnetic barrier and are turned into low-field seekers by the spin-flipper to be stored magnetically.

In the case of material filling, the reflection capabilities of suitable materials for UCNs can be used during the filling procedure to accumulate density in the trap; subsequently the magnetic trap is created by ramping up coils.[c] Without pre-polarizing the UCNs, the ratio of high-field seekers to low-field seekers will still be relatively large after the magnetic field is at

---

[c]A permanent magnet trap can not be used in this mode.

its nominal value. The storage lifetime of high-field seekers is much longer than in the magnetic filling case, since the material walls surrounding the magnetic trap have to be UCN reflectors. To reduce their fraction down to an acceptable level, the trap may be cleaned of high-field seekers with a carefully positioned or movable absorber.[19,20]

The neutron lifetime experiment PENeLOPE (Precision Experiment on the Neutron Lifetime Operating with Proton Extraction) will apply both methods and shall be discussed in more detail in the following section.

4. The Polarization System for PENeLOPE

As most next generation neutron lifetime experiments, PENeLOPE is a magneto-gravitational trap: 28 superconducting magnets create a low-field region surrounded by a high-field region to confine ultra-cold neutrons, see Fig. 1. Due to gravity, the top of the trap can stay open, facilitating extraction of the charged decay particles.[20]

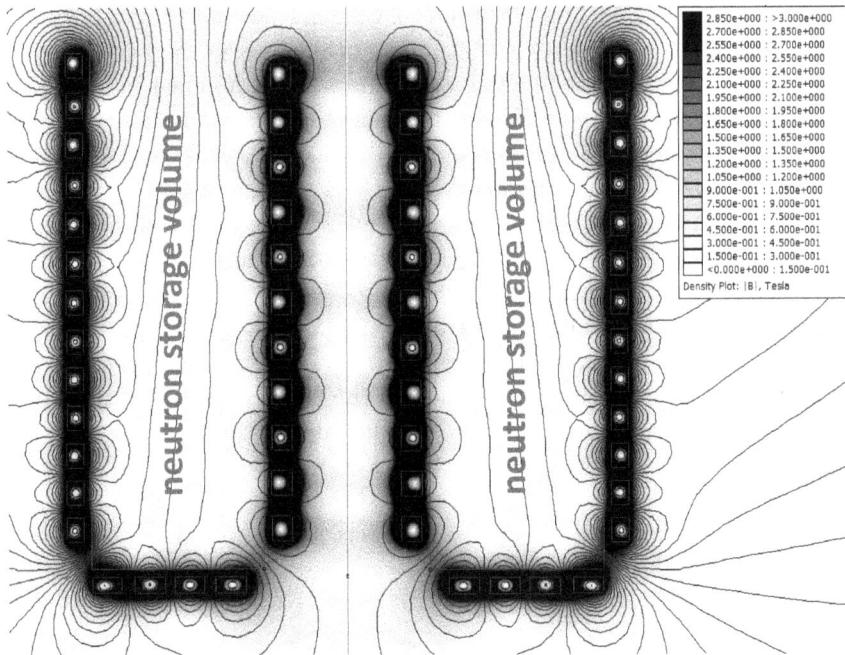

Fig. 1. Flux density (gray shading) and field lines of the PENeLOPE magnet, vertical cut.

160

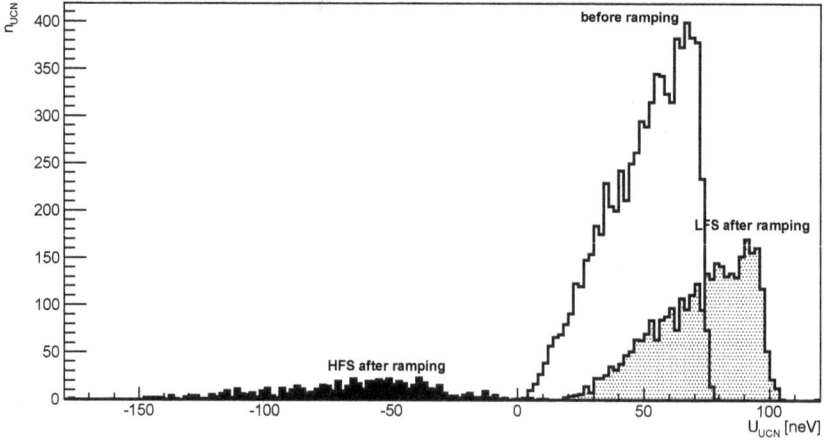

Fig. 2. UCN energy spectra from Monte Carlo simulations before and after ramping up the magnet of PENeLOPE. The reference potential $U_{UCN} = 0$ is set to the bottom of the storage volume.

While the UCNs are accumulated from the UCN source, all superconducting coils have to be turned off;[d] UCNs are held in the trap via an electro-polished stainless steel vessel, lining the inside of the neutron storage volume. The superconducting coils are specially designed to allow fast ramping in less than 100 s. During this ramp up, low- and high-field seekers start to experience forces in opposite directions. The effect of this is shown in Figs. 2 and 3: the low-field seekers are starting to be confined in the low-flux-density region, hence are not experiencing high-flux densities as the high-field seekers do. They are heated by around 25 neV almost uniformly. The HFS on the other hand are accelerated towards the high flux-densities at the walls of the trap, which are potential depressions for this UCN species. They are confined spatially between the low field in the center of the neutron storage volume and the Fermi potential of the material walls. In these regions, the change of flux density during ramping is up to around 4 T, much larger than for the LFS. As a consequence, the cooling effect on the HFS is much bigger than the heating effect for the LFS and they end up at negative energies with respect to the bottom of the storage volume, see Fig. 2.

---

[d]Partially shutting down the magnet to create an opening for low-field seekers is not recommended for highly coupled superconducting coils. Additionally, the filling orifice created would have been small and would have a significant magnetic barrier, thus reducing the effective trap depth of PENeLOPE.

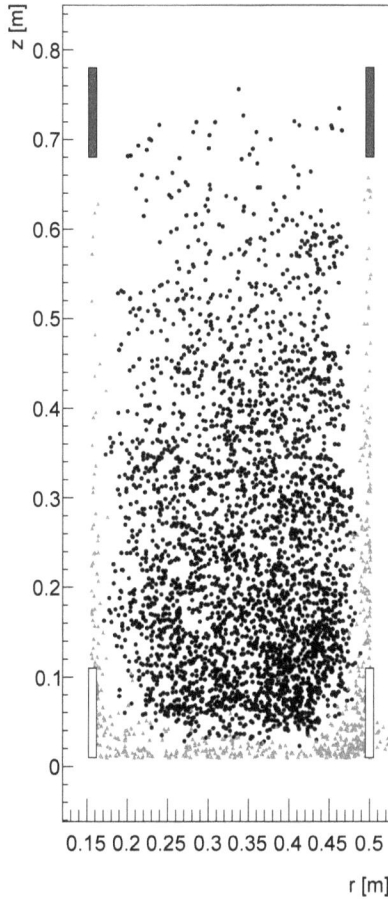

Fig. 3. Scatter plot of low-field seekers (black dots) and high-field seekers (gray triangles), showing their spatial distribution right after ramping up the magnets of PENeLOPE (Monte Carlo simulation). The upper absorber ring positions are marked with black rectangles, the lower with white ones.

Figure 3 shows the resulting spatial distribution of low- and high-field seekers after ramping: there is almost no overlap between the two spin alignments. Due to this, an absorber can be put into the regions that are only populated by HFS to remove them from the trap before starting the measurement.[e]

---

[e]The method of filling in high-field seekers and performing the spin flip in-situ is not applicable due to the complicated topology of both, the magnetic field and the cryostat.

The experiment procedure of PENeLOPE requires two absorber rings to remove UCNs with an energy exceeding the magnetic trap depth of PENe-LOPE (115 neV). Because of the LFS heating, these polyethylene rings will be placed 68 cm above the bottom of the storage volume, corresponding to a lower potential of around 70 neV in the gravitational field (cf. Fig. 3). They are hidden from the magnetically stored spin species after ramping due to the magnetic field barrier, but effectively remove UCNs with energies > 80 neV from the trap (see Fig. 2), hence avoiding the so called marginally trapped neutrons.[21]

Since the high-field seekers have very low energies in the magneto-gravitational potential after ramping, the absorbers are situated outside of the accessible phase-space for a majority of HFS, leading to a very long cleaning time. The main loss mechanism for them is absorption or up-scattering at the stainless steel walls, hence, reaching an acceptably low fraction $n_{HFS}/n_{LFS} < 10^{-4}$ would take up to 1000 s. The actual measurement can only start after this cleaning time; especially the decay particle measurement is very sensitive to neutrons with shorter storage lifetimes. The precision goal of PENeLOPE is $\Delta\tau_n < 0.1$ s. The systematic decrease of the neutron lifetime measured with decay particles due to high-field seeker contamination drops below 0.03 s, when the fraction n_{HFS}/n_{LFS} is below $2 \cdot 10^{-4}$.[f]

Monte Carlo simulations show that moving the absorber down to the bottom of the storage volume (cf. Fig. 3) decreases the required cleaning time to around 400 s, see Fig. 4. To achieve that, the two absorber rings will be suspended from the vacuum lid on top of the neutron storage volume. The rings will be moved down to the bottom of the storage volume after the current in the superconducting magnet has reached its nominal value; Linear drives connected to their vacuum feedthroughs will enable short moving times of 5 to 10 s. As soon as an acceptable fraction n_{HFS}/n_{LFS} is reached, they will be moved back up.

Only with additional pre-polarization, one can shorten the cleaning period further to around 150 s; this increases the counting statistics by about 30%, because fewer neutrons decay during the waiting period. The plot in Fig. 4 assumed a pre-polarization of 80%; as shown in Fig. 5, this will be achieved by using an iron-coated aluminum foil in the guide system, magnetically saturated by a surrounding Halbach array of permanent magnets.

---

[f]The UCN counting measurement is much less sensitive; for a shortest storage time of 1000 s, the systematic shift is less than 0.03 s even up to $n_{HFS}/n_{LFS} = 10^{-2}$.

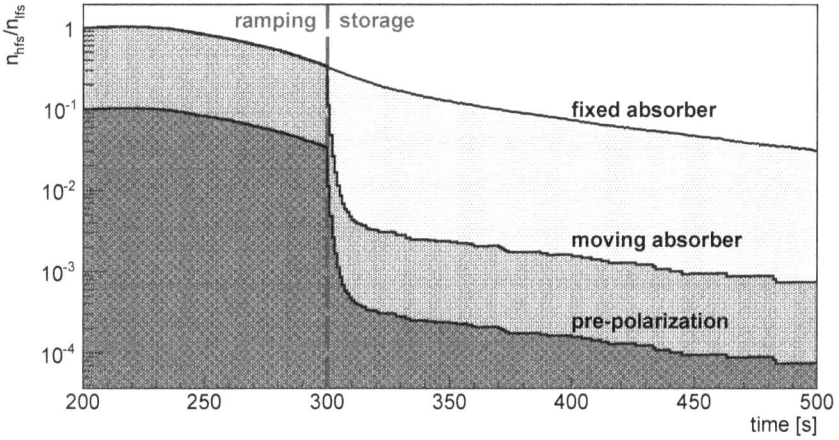

Fig. 4. Monte Carlo simulation of the high-field seeker cleaning in PENeLOPE. Top histogram: Polyethylene rings are fixed at a height of 0.68 m above the bottom of the storage volume. Middle histogram: Absorber ring is moved down to the bottom after ramping. Bottom histogram: Additionally to moving the absorber, pre-polarized UCN (80%) are filled into the trap.

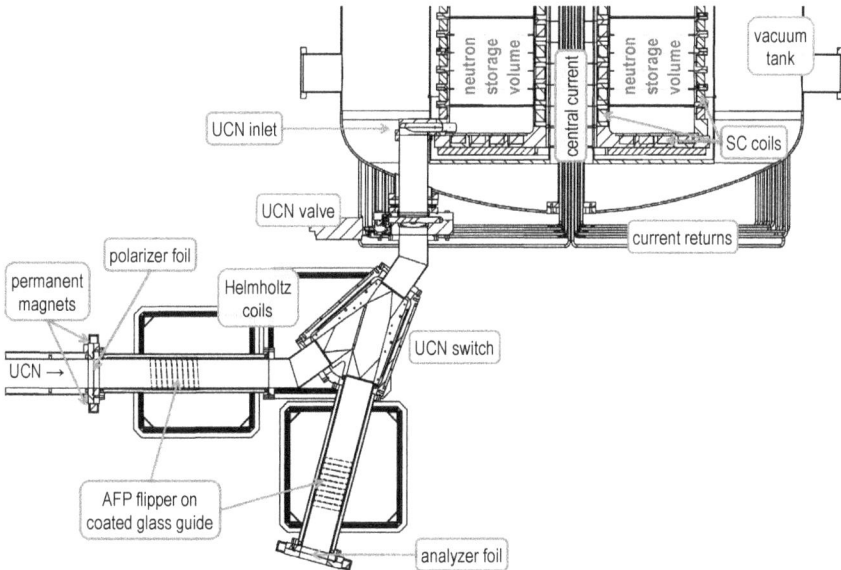

Fig. 5. Drawing of the polarization system of PENeLOPE.

Polarizations of 90 to 95% were demonstrated with this approach.[22] The transverse polarizer field of at least 10 mT smoothly transitions to a transverse guiding field of approx. 1 mT using rectangular Helmholtz coils. In this gradient, an adiabatic fast passage (AFP) flipper rotates the neutron spin by 180° with respect to the field. The flipper consists of a radio-frequency solenoid mounted onto a NiMo-coated glass guide and produces an axial field of approx. 0.5 mT (RMS) with a frequency of 30 to 60 kHz. Spin flip efficiencies > 97% have been demonstrated.[13] Additional Helmholtz coils produce a spin guiding field (ca 1 mT) between the polarizer and the storage volume to reduce depolarization. The guide system in this region consists of stainless steel and NiMo-coated parts; these materials have an average depolarization probability per wall bounce of $< 10^{-5}$.[23] The maximum number of wall reflections for the UCN filled into PENeLOPE is 3000, leading to a UCN polarization of > 84% after ramping. Inside the storage volume a current of up to 20 kA at the rotational axis of PENeLOPE produces both the spin holding field during filling and ramping and also prevents spin-flips in the low and zero field regions of the magnetic trap of PENeLOPE during storage. After the storage period an optional second flipper and polarizer in front of the detector allows polarization analysis for systematic studies of the polarization system.

5. Conclusion

Most next generation neutron lifetime experiments aim at a precision of $\Delta\tau_n < 0.1s$ and will store polarized ultra-cold neutrons in a magnetic trap. The necessary polarization of the stored UCN depends on the measurement technique: counting the decay particles emitted during neutron decay imposes a much more stringent requirement than UCN counting. For PENeLOPE, where the shortest UCN storage time will be greater than 500 s, the initial polarization for UCN counting should be greater than 99%, whereas is has to exceed 99.98% for decay particle counting.[g] To accomplish this, a two stage approach is necessary. The UCN will be polarized in the guide system before entering the neutron trap. With current techniques, it is not possible to achieve a 99.98% polarization, mainly due to depolarization during UCN transport. Hence, the excess high-field seekers will be removed using carefully placed neutron absorbers.

---

[g]All the high-field seekers lost during the shortest foreseen storage time (> 500 s) and before emptying the trap will not contribute to the false effect which lowers the neutron lifetime result.

PENeLOPE is suppported by the Deutsche Forschungsgemeinschaft (DFG), the DFG cluster of excellence 'Origin and Structure of the Universe' and the Maier-Leibnitz-Laboratorium. Thanks go to A. Frei for the UCN switch and T. Lauer for guide coating.

References

1. D. Dubbers and M. G. Schmidt, *Rev. Mod. Phys.* **83**, 1111 (2011).
2. B. Märkisch *et al.*, *Nucl. Instr. Meth. A* **611**, 216 (2009).
3. M. P. Mendenhall *et al.*, *Phys. Rev. C* **87**, p. 032501 (2013).
4. S. Baessler *et al.*, *Eur. Phys. J. A* **38**, 17 (2008).
5. M. Antonelli *et al.*, *Eur. Phys. J. C* **69**, 399 (2010).
6. S. Arzumanov *et al.*, *Phys. Lett. B* **483**, 15 (2000).
7. A. P. Serebrov *et al.*, *Phys. Rev. C* **78**, p. 035505(Sep 2008).
8. Arzumanov *et al.*, *JETP Lett.* **95**, 224 (2012).
9. A. Pichlmaier *et al.*, *Phys. Lett. B* **693**, 221 (2010).
10. J. Beringer *et al.*, *Phys. Rev. D* **86**, p. 010001(Jul 2012).
11. S. Paul, *Nucl. Instr. Meth. A* **611**, 157 (2009).
12. A. T. Holley *et al.*, *Rev. Sci. Instrum.* **83**, p. 073505 (2012).
13. P. Geltenbort *et al.*, *Nucl. Instr. Meth. A* **608**, 132 (2009).
14. F. Atchison *et al.*, *Phys. Rev. C* **76**, p. 044001(Oct 2007).
15. A. Serebrov *et al.*, *Phys. Lett. A* **313**, 373 (2003).
16. V. Ezhov *et al.*, *Nucl. Instr. Meth. A* **611**, 167 (2009).
17. P. Walstrom *et al.*, *Nucl. Instr. Meth. A* **599**, 82 (2009).
18. C. O'Shaughnessy *et al.*, *Nucl. Instr. Meth. A* **611**, 171 (2009).
19. K. Leung and O. Zimmer, *Nucl. Instr. Meth. A* **611**, 181 (2009).
20. S. Materne *et al.*, *Nucl. Instr. Meth. A* **611**, 176 (2009).
21. R. Picker *et al.*, *Nucl. Instr. Meth. A* **611**, 297 (2009).
22. R. Herdin *et al.*, *Nucl. Instr. Meth.* **148**, 353 (1978).
23. M. Daum *et al.*, *Phys. Lett. B* **704**, 456 (2011).

Summary of Workshop on Next Generation Experiments on the Neutron Lifetime

DIRK DUBBERS

University of Heidelberg
Heidelberg, Germany

KRISHNA S. KUMAR

University of Massachusetts
Amherst, USA

JOHN M. PENDLEBURY

University of Sussex
Brighton, UK

This workshop addressed the state of present day experiments seeking to measure the neutron lifetime. Measurements made using cold neutron beams as well as using trapped ultra-cold neutrons were reviewed. A number of groups discussed prospects for improvements in both types of experiments. Focus was given to understanding the systematic uncertainties that must be addressed to make progress in the precision of the measured lifetime. In this paper we summarize the major conclusion from this workshop.

Semileptonic weak processes between leptons and quarks play a dominant role in the rapidly evolving fields of particle, nuclear and neutrino physics, astrophysics and cosmology. Today the measured neutron lifetime is still the only source to derive the various *semileptonic* charged weak interaction cross sections needed in these fields. The neutron lifetime is now at $\tau_n = 880.1 \pm 1.1$ s[1] (Particle Data Group) but in-beam and in-trap lifetime measurements differ by up to one percent, with corresponding uncertainties in calculated weak interaction rates. In contrast, the *purely leptonic* cross sections, as derived from the measured muon lifetime, are now known to one part in a million, and the resulting errors can be neglected in practically all applications.

While, over the years, the error of the neutron lifetime has diminished by a factor of 200, progress was slow, the error being divided by 2 or 3 every ten years. Furthermore, at all times the lifetime error was underestimated by a factor of three, and it is not excluded that this is still the case. Our ultimate aim should be that, when dealing with semileptonic processes, one should be able to safely neglect all uncertainties from the neutron lifetime, in the same way as in the purely leptonic case. This requires the lifetime to be known to an uncertainty well below 1 s. Indeed, there are compelling physics reasons that motivate efforts to aim for an uncertainty approaching 0.1 s. This aim requires a continuing international effort on neutron lifetime measurements.

For *trapped UCN*, the standard procedure is to register the exponential decrease in the number of surviving UCN for increasing storage times, with successive loadings of the trap.

Systematic errors in UCN *magnetic storage* experiments to determine the neutron beta decay lifetime are:

A. Mechanisms other than beta decay that can remove stored UCN from the trap
   1) Majorana spin flips
   2) Insufficient initial spectral cleaning to remove 'marginally trapped' UCN
   3) Gas scattering of the UCN due to H_2 molecules
   4) UCN warming due to mechanical vibrations of the trap
   5) UCN warming due to noise on the B fields
   6) Uncontrolled variations in the initial UCN loading
   7) Incomplete emptying of the trap
   8) Leaks through any low magnitude holes in the B field

B. UCN detection problems
   1) Efficiency changes due to UCN warming in storage
   2) Inclusion of 'marginally trapped' UCN due to insufficient spectral cleaning
   3) General background counts
   4) UCN capture induced background near the detector
   5) Detector dead time losses
   6) Errors and uncertainties in the time of the detection
   7) When detecting decay products, build-up times and particle losses for the products

In the latest lifetime experiments using stored UCN, the individual errors assigned to these points are all well below 1 s, with total errors of order 1 s.

For *material traps*, leaks are dominated by inelastic 'up-scattering'. In the case of liquid surfaces there is also quasi-elastic scattering of UCN from thermally excited surface capillary waves. Also observed for solid walls there is quasi-elastic 'anomalous heating' at the level of about 10^{-5} per reflection, the latter being of unknown origin.

Concerning statistics, at present the typical number of stored UCN in one filling is of order 10^4 (typical neutron densities being several UCN/cm^3 in volumes of several liters). With effective cycling periods of about an hour, the integrated number of UCN stored in a lifetime experiment of several months duration is of order 10^7. For a lifetime measurement to 0.1 s with one fixed set of parameters, one finds that 10^9 stored UCN are needed. Thus, when the parameters are varied to study the systematic effects for a 0.1 s lifetime error one realistically should foresee the need for an integrated total of 10^{10} stored UCN.

We are only aware of one neutron lifetime experiment using stored UCN, which is in a data-taking phase. This is the Ezhov experiment using permanent magnets, which runs at the ILL on Level D. The UCN storage volume is 12 liters but the density is only about 1 UCN/cm^3. A further experiment called 'HOPE' uses a superconducting multi-pole magnet and will have a UCN storage volume of about 2 liters and is at the initial tests stage. They hope, in future, to store a UCN with a density of about 50 UCN cm^{-3}. The UCN will come from a dedicated source on the ILL H172 beam. A third experiment called PENeLOPE is in the early stages of design and construction at TU Munich. This has an annular UCN storage volume of 400 liters bounded by superconducting multipole fields. It may be a few years yet before it is ready to take data.

For *in-beam* experiments, main errors enter via the effective number of neutrons in the active beam volume. At present, the (n, α) cross section and thickness of the thin foils used for flux calibration, the solid angle of α detection, and the statistics, each contribute an error of about 1 s to the lifetime result. As discussed later, there is significant potential to improve on these errors.

These error sources, together with the strong scatter seen in earlier lifetime results, make it advisable to carry out several independent lifetime experiments, preferably with different sources of error. Worldwide, about one-half dozen neutron lifetime experiments are running or are under construction.

When, in the decade to come, another factor of three or more suppression of the error in τ_n is envisaged, other small effects will come to the forefront. As was shown in earlier storage experiments, even the movement of an UCN valve can change the UCN spectrum and with it the storage time. In magnetic traps, small cracks in permanent magnets may disturb the nearby magnetic field. Even UCN scattering on residual atoms in the vacuum may require a correction. On the other hand, additional experimental tools will become available, like the detection of decay particles in order to follow exponential decay separately during each filling cycle. Furthermore, the advent of new powerful UCN sources will diminish the statistical error, and at the same time permit more detailed studies of systematics.

At the Santa Fe workshop, three US instruments to measure the neutron lifetime were presented, two at the NIST reactor neutron source, and one at the LANL accelerator neutron source.

Over the years, the NIST *in-beam* experiment has produced lifetime values that are significantly higher than the more recent UCN in-trap lifetime results. In the LANL workshop, the collaboration convincingly presented a series of new inter-calibration procedures that will permit substantial error reduction. Given that the advancement of calibration methods is at the heart of NIST's activities, and given that alternatives in neutron lifetime measurement methods with entirely different sources of error are highly welcome, we are of the opinion that this in-beam program clearly deserves further strong support.

The NIST *in-trap* experiment works with an in-situ 'super-thermal' UCN source. The ^{4}He UCN converter at the same time serves as a scintillator for the electrons from neutron decay. This approach is very intriguing, but, in spite of heroic efforts, it evidently turned out to be too difficult to realize, due to insurmountable background and luminescence problems. Two years ago, the internal structure of the trap was damaged, and the collaboration seems to have lost confidence in a successful continuation of the project. It is therefore doubtful whether this project can be resurrected.

The UCN lifetime project at LANSCE uses a hybrid magnet configuration, consisting of a 'bowl' made of permanent magnets ('Halbach array') for magneto-gravitational trapping, plus a superconducting holding field that ensures adiabatic following of the magnetic field by the UCN spins. This last field may also permit magnetic guiding of decay protons to a detector. UCN counting with low background can begin within a few seconds of initiation, using a vanadium sheet activation method that profits from the vanadium negative Fermi potential (the few seconds are due to the delay needed while the UCN are all absorbed by the vanadium). The project is well underway, in simulation, design, and construction. The number of stored UCN per filling expected is more than three orders of magnitude higher than in the previous experiments that we mentioned above. Its effective UCN storage volume is approximately 700 liters – similar to the PENeLOPE magnet but with a very different shape and likely to be operational some years earlier. This is a 60-fold increase in volume compared with that of the Ezhov experiment. With this in mind, it is clear that the statistics that can be reached at LANSCE are sufficient for the very demanding goal of 0.1 s uncertainty in the neutron lifetime, particularly if the proposers can arrange their instrument to be closer to the UCN source, and provided that their proposed UCN detection methods work properly.

An intermediate aim of UCNτ at LANSCE should be a reliable 1 second measurement of the neutron lifetime, with a good potential to do better once more powerful UCN sources are available. It would be highly desirable if this experiment could proceed to make progress following the strategies that were presented. Should the project at LANSCE make the first measurements that show the potential to reach well below 1 second with the limited UCN density available, the experience gained and the technical problems encountered and solved along the way would allow the proponents to make a powerful case a few years down the road to obtain substantial new funding that would be required to reach the very demanding goal of 0.1 s. It is also hoped that the proponents will actively recruit new collaborators both within the US and internationally once a realistic design and funding proposal begins to develop to reach this goal.

The Santa Fe meeting ended after hosting a discussion on blind data analysis in the context of these experiments. In general, blind data analysis is a valuable tool to avoid bias, although cases may be conceived where intermediate results are needed to decide on the further evaluation procedure. After all, the best rules do not suffice to replace good character.

Reference

1. J. Beringer *et al.*, Physical Review D **86**, 010001 (2012).

Color Plate 1

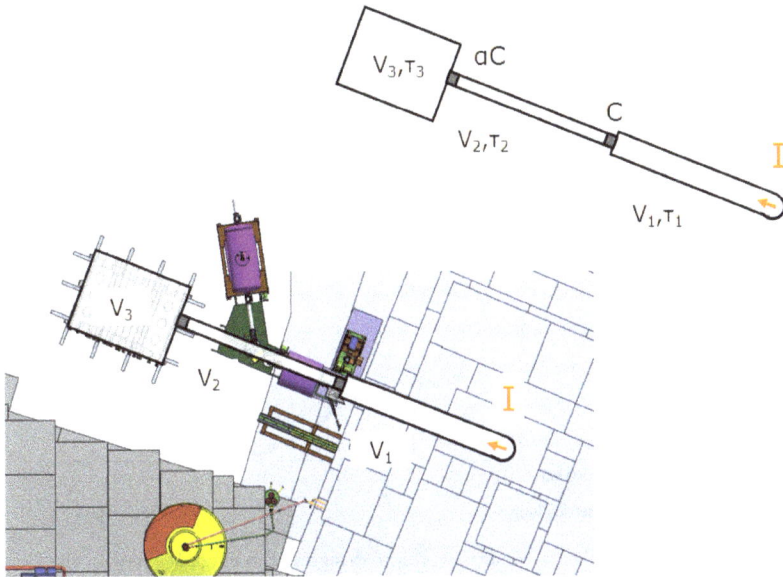

Fig. 1. Model Geometry for the Area B source coupled to the UCNτ experiment. (See page 32)

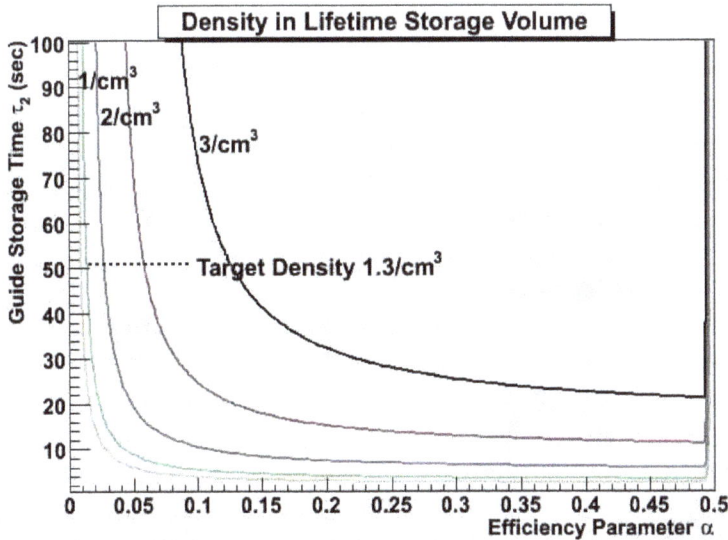

Fig. 2. Model results for the LANL Area B source, where we use approximate values $I_e = 1.7 \times 10^4/\text{s}$, $C = 2300 \text{ cm}^3/\text{s}$, $\tau_1 = 30\text{s}$ and $\tau_2 = 880\text{s}$. (See page 33)

Color Plate 2

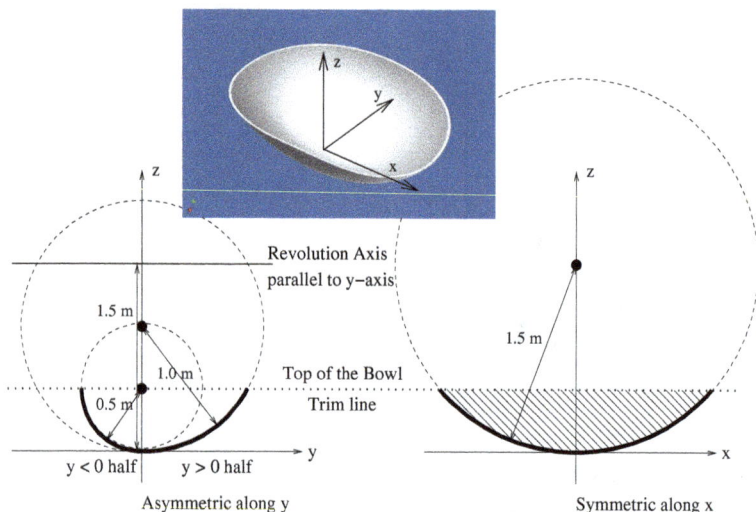

Fig. 1. A schematic of the asymmetric toroidal design of the magnet array. In the global coordinate system, the x axis is along the symmetric axis of the toroid (with the major radius $R = 1.0$ m), while the y axis is along the asymmetric axis (with different minor radii, $r_R = 1.0$ m and $r_L = 0.5$ m) and the z axis points up vertically. The thick solid curves represent the inner-surface profile of the Halbach array. (See page 46)

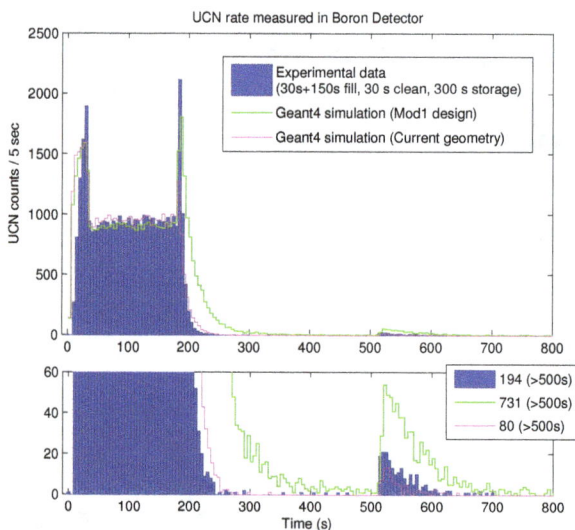

Fig. 2. A comparison between the GEANT4 simulated timing spectra and the measurement in the "fill & dump" mode (blue histogram: 180 s filling, 30 s cleaning and 300 s storage.). The magenta curve is the simulated result using current geometry, while the green curve is the predicted enhancement with a MOD1 design, which is used to increase the UCN loading efficiency into the trap. (See page 47)

Color Plate 3

Location of Absorption

Asymmetric y with field ripples

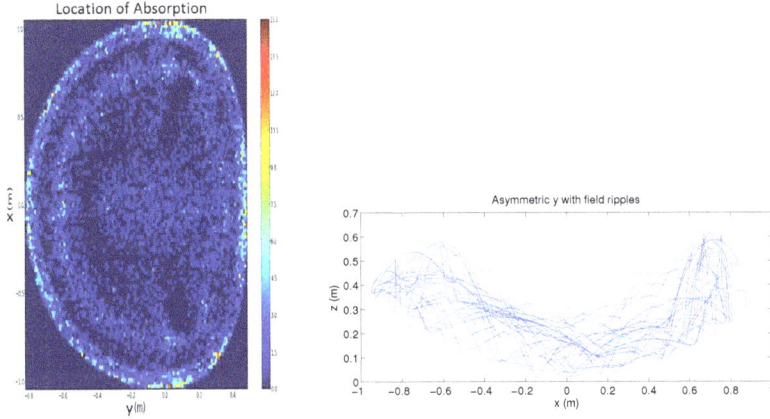

Fig. 3. *Left*: A 2-D histogram of the absorption events of the high-energy UCN (with $E = V_0 + 6$ neV) on a UCN absorber, fully covering the top of the trap. *Right*: A example 3-D trajectory of a high-energy UCN inside the trap, numerically integrated using a simplectic integrator of the full Halbach magnetic potential (on the asymmetric toroid with field ripples). (See page 48)

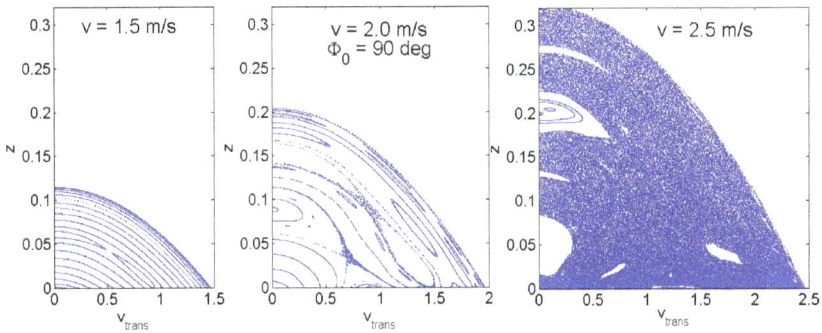

Fig. 5. Poincaré maps plotting v_θ and z at the points of landing on the surface of the trap bowl. The geometry of the trap is the same degenerate sphere used in Fig.4. The transverse velocity is $v_{trans} = v_\theta$. (See page 52)

Color Plate 4

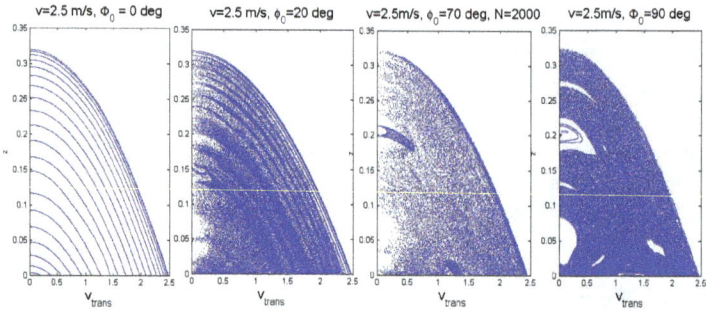

Fig. 6. Poincaré maps of the UCN dynamics in a vertical toroidal trap with a major radius $R = 1.0$ m and a minor radius $r = 0.5$ m. The neutrons are of the same velocity $v = 2.5$ m/s starting at the bottom of the bowl. The incoming polar angle θ is varied the same way as in Fig.4, but the azimuthal angle ϕ is varied from 0 (left graph) to 90° (right graph). The transverse velocity is $v_{trans} = \sqrt{v_\theta^2 + v_\phi^2}$. (See page 53)

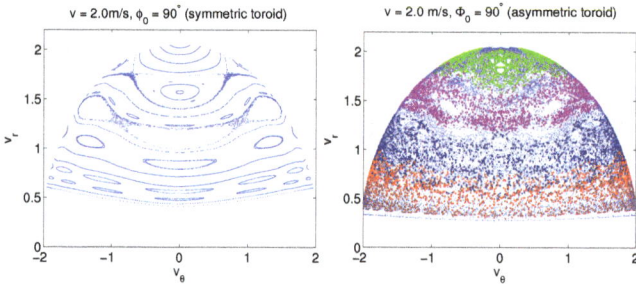

Fig. 7. Poincaré maps of the UCN dynamics in a symmetric toroidal trap (*left figure*) and an asymmetric toroidal trap (*right figure*), with $R = 1.0$ m, $r_R = 1.0$ m and $r_L = 0.5$ m. These motions are confined to $\phi = 90°$ on a symmetry plane. On the right figure, groups of 4 different θ_0 were highlighted in colors to illustrate the corresponding region of stochasticity. (See page 54)

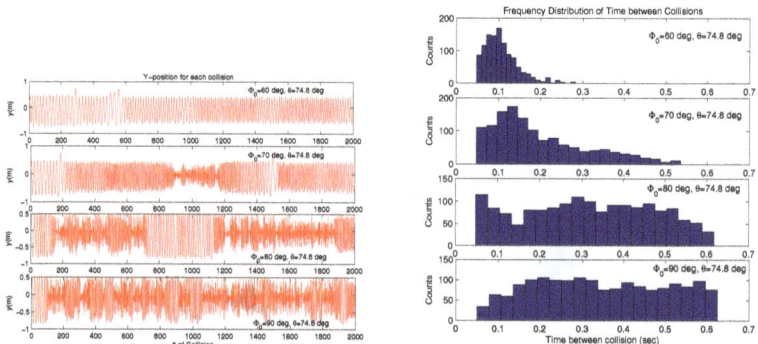

Fig. 8. History of the collisional position (*left*) and frequency distributions of time between collisions (*right*). (See page 55)

Color Plate 5

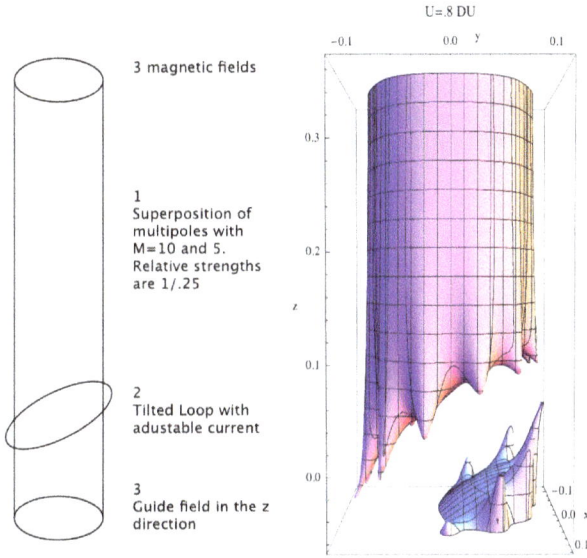

3 magnetic fields

1
Superposition of
multipoles with
M=10 and 5.
Relative strengths
are 1/.25

2
Tilted Loop with
adustable current

3
Guide field in the z
direction

U=.8 DU

Fig. 1. The principle geometry of the trap is shown in the left. At some energy U_{max}, the upper and lower equipotential surfaces coalesce and neutrons are not trapped. On the right is the equipotential surface for neutrons with $E_n = 0.8\ DU$. Only the lower quarter of the equipotential surface is shown. Less than 10^{-3} of the orbits with $E_n < 0.8\ DU$ are regular with the loop tilted to 30 degrees. (See page 60)

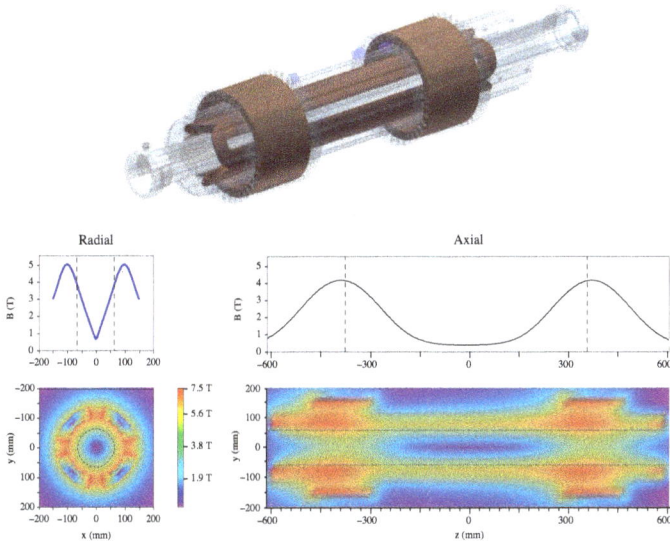

Fig. 1. Schematic of NIST magnetic trap. (See page 67)

Color Plate 6

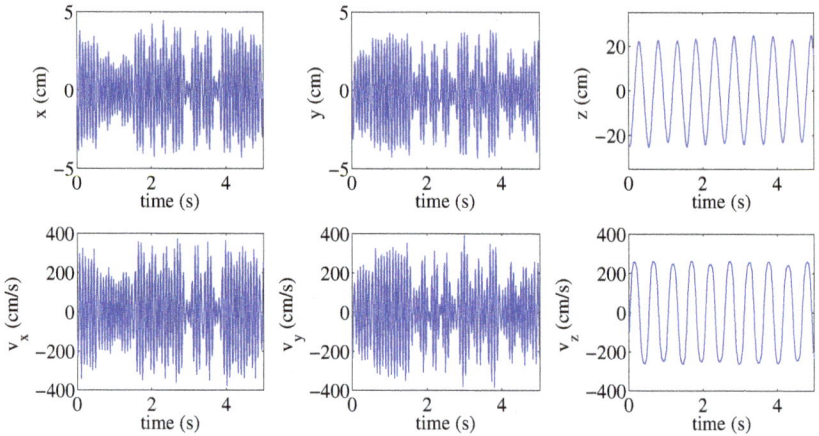

Fig. 2. Sample trajectory determined by symplectic integration method. (See page 68)

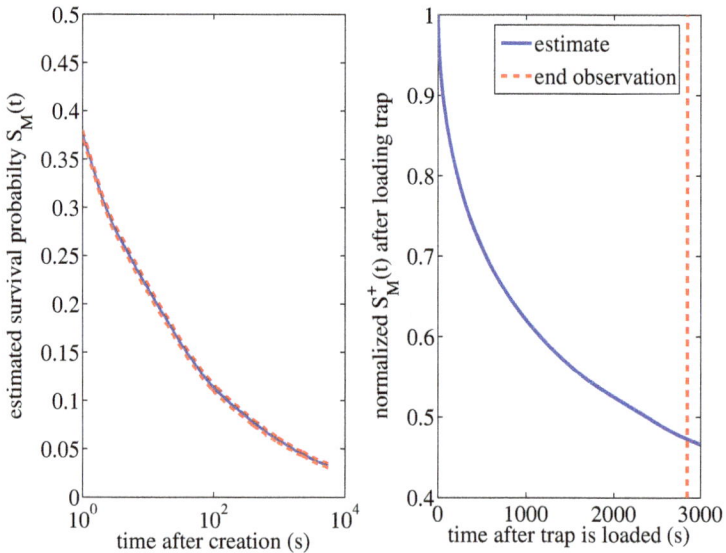

Fig. 3. Left: Monte Carlo estimate of the survival probability associated with the wall loss mechanism for an ensemble of above threshold UCNs and associated ±-1 standard error bands. Right: conditional survival probability for wall loss mechanism for UCNs that survive until loading stage is completed. In this simulation study, UCNs are lost by either β−decay or the marginally trapped loss mechanism. In this simulation study $\tau_n = 880.1$ s. (See page 71)

Color Plate 7

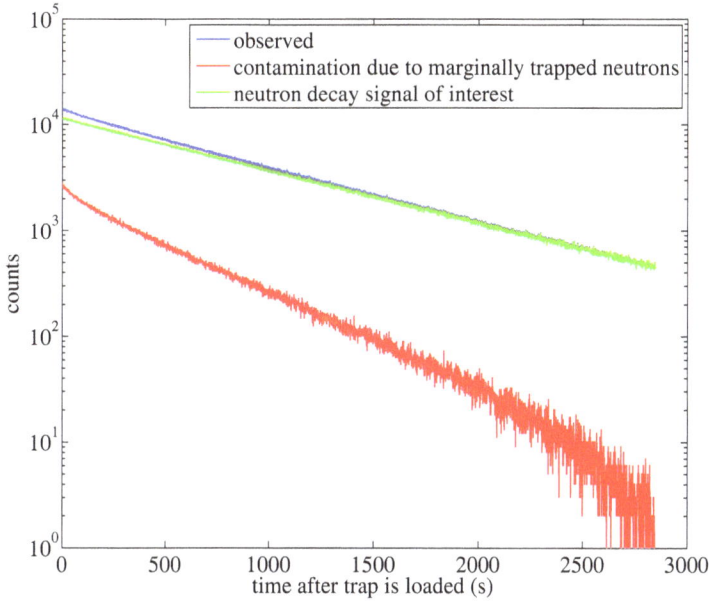

Fig. 4. Simulated β−decay data. Neutron signal of interest corresponds to Eq. 19. The contamination signal corresponds to Eq. 20. In this simulation study $\tau_n = 880.1$ s. (See page 73)

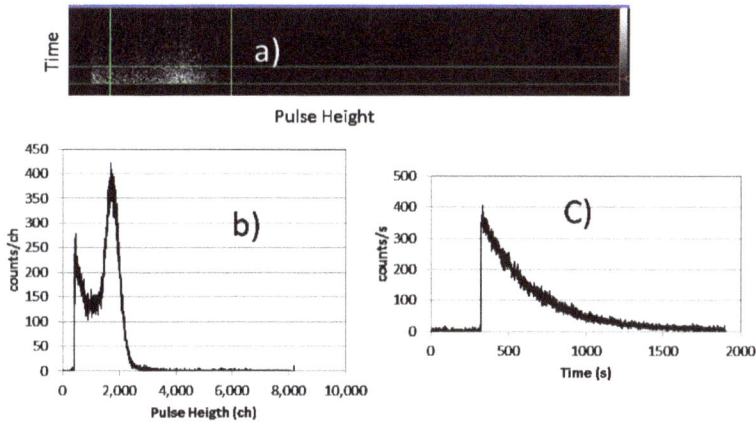

Fig. 5. Data taken to test vanadium activation for counting UCN. a) is a two dimensional plot of γ pulse height vs. time for β−γ coincidences. b) Shows the coincident γ-ray pulse height distribution summed over the region marked by the horizontal green lines. The 1.43 MeV γ-ray peak is evident. c) Shows the coincident γ-ray count rate, summed over the region marked by the vertical green lines. (See page 119)

Color Plate 8

Fig. 1. The neutron lifetime apparatus on the NCNR beamline. (See page 123)

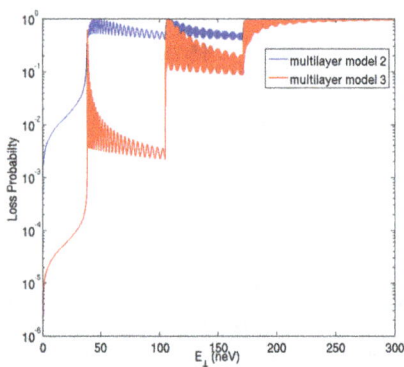

Fig. 3. The multi-layer models for the material wall potential used in modelling above-threshold neutrons. (See page 127)

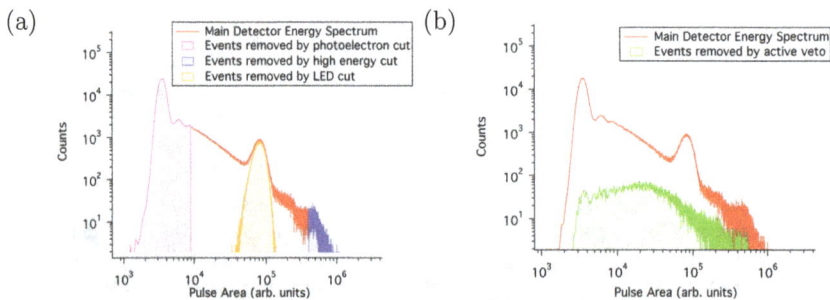

Fig. 4. (a) Application of threshold and pulser event cuts. (b) Removal of cosmic ray events. (See page 130)

Color Plate 9

(a)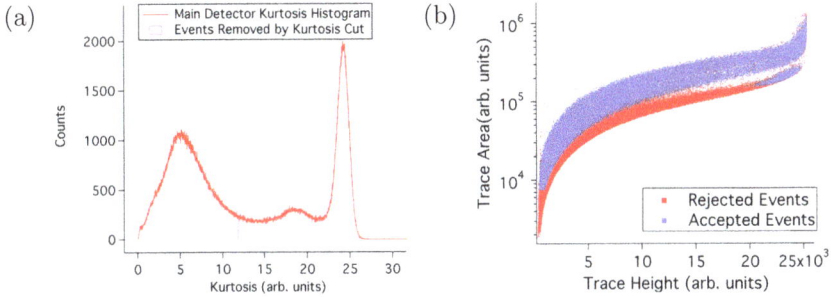

(b)

Fig. 5.   (a) Events histogramed by gaussian kurtosis and illustrating pulse shape analysis of the data. (b) Two-dimensional plot of pulse area vs height; events in the lower band (associated with the plastic) not removed by the pulse shape cut are removed later with specific pulse height cuts. (See page 131)

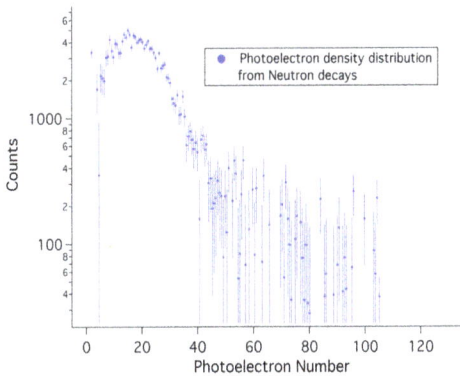

Fig. 6.   Neutron trapping data after all background cuts from above (single channel). (See page 132)

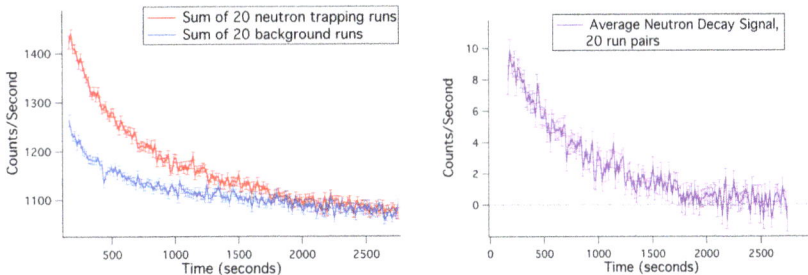

Fig. 7.   Neutron trapping and non-trapping data after all background cuts (left) and the neutron decay signal after subtraction of trapping and non-trapping data runs (right). (See page 132)

182

Color Plate 10

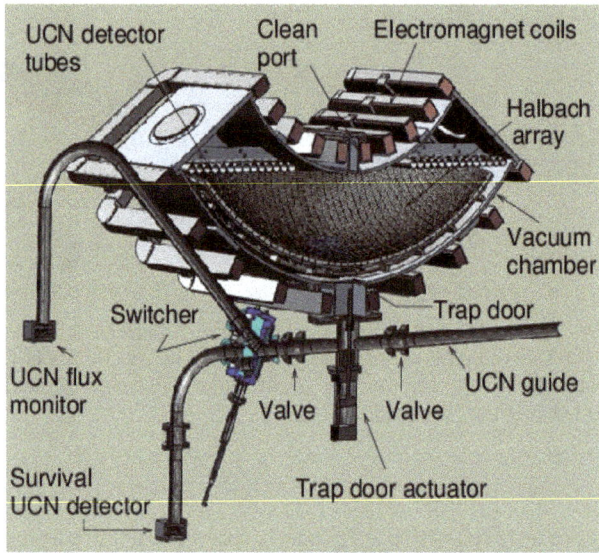

Fig. 1.   Layout of the UCNτ neutron lifetime experiment. (See page 137)

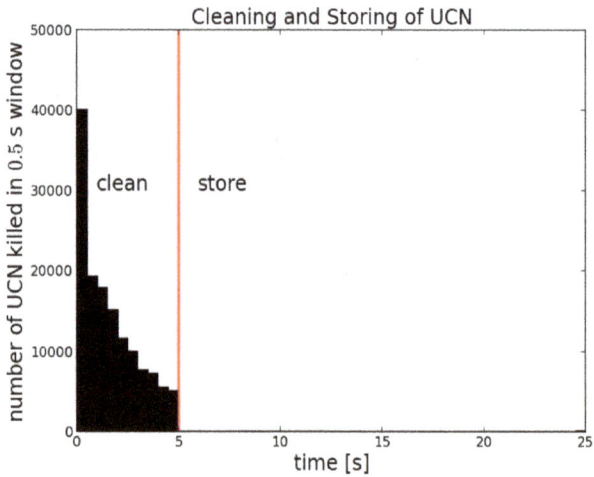

Fig. 3.   Simulated cleaning of quasi-bound neutrons from the trap. The UCN source can be used to make data-driven predictions.[8] As an example, UCN are created upstream of the experiment with an experimentally determined initial velocity spectrum. From the time of flight of the UCN, we can predict the loading time of the trap for different guide configurations, and source fluctuations can be included. (See page 140)

www.ingramcontent.com/pod-product-compliance
Lightning Source LLC
Chambersburg PA
CBHW050628190326
41458CB00008B/2180